Glossary of
Molecular Biology

Glossary of Molecular Biology

A. EVANS

LONDON
BUTTERWORTHS

THE BUTTERWORTH GROUP

ENGLAND
Butterworth & Co (Publishers) Ltd
London: 88 Kingsway, WC2B 6AB
AUSTRALIA
Butterworths Pty Ltd
Sydney: 586 Pacific Highway, NSW 2067
Melbourne: 343 Little Collins Street, 3000
Brisbane: 240 Queen Street, 4000
CANADA
Butterworth & Co (Canada) Ltd
Ontario: 2265 Midland Avenue, Scarborough
NEW ZEALAND
Butterworths of New Zealand Ltd
Wellington: 26–28 Waring Taylor Street, 1
SOUTH AFRICA
Butterworth & Co (South Africa) (Pty) Ltd
Durban: 152–154 Gale Street

First published 1974

© Butterworth & Co (Publishers) Ltd 1974
ISBN 0 408 70640 6 Standard
0 408 70641 4 Limp

Printed in England by
The Camelot Press Ltd, Southampton

Preface

In a glossary, above all literature concerned with this subject, 'molecular biology' should surely be defined. The phrase clearly cannot be taken at face value, for if all biological matter is molecular, is not all biology molecular? Indeed, it has been written before that not even those who are known as molecular biologists would be unanimous in defining their titles. W. T. Astbury, probably the first to give a definition of the spheres of molecular biology and the first to use the term, said that it 'implies not so much a technique as an approach, an approach from the viewpoint of the so-called basic sciences with the leading idea of searching below the large-scale manifestations of biology for the corresponding molecular plan. It is concerned particularly with the *forms* of biological molecules, and with the evolution, exploitation and ramification of these forms in the ascent to higher and higher levels of organisation. Molecular biology is predominantly three-dimensional and structural—which does not mean, however, that it is merely a refinement of morphology. It must at the same time inquire into genesis and function.'

A look at the contents of any journal, review or book concerned with molecular biology today will reveal that this original concept has been greatly extended if not overtaken. However, in their search for molecular explanations of the features that characterise living matter, molecular biologists are as one in as much as the macromolecules unique to cells are the molecules with which they are mainly involved. The diversity of specialities within which this involvement may arise, including cytology, genetics, immunology, microbiology, virology, has in fact made it difficult to decide in the compilation of this glossary how many of the terms more closely pertinent to the specialities should be included, a difficulty not lessened by all of these specialities already having very sizeable vocabularies of their own. The inclusion of such terms has therefore been restricted to the minimum necessary to cover their use within the *Glossary* itself. By indicating in the entries, where relevant, the type of organisms used in the background work I hope to have met the criticism that in reporting the findings of molecular biologists it often seems to be held that what is true for *Escherichia coli* is also true for *Homo sapiens*.

The *Glossary* has been prepared for the use of those reading the specialised literature of molecular biology (or even the fashionable digested reviews of such literature), who are chemists, as well as for those whose background is primarily biological or clinical. It is hoped that the *Glossary* will also be found helpful by students in all relevant disciplines.

A. EVANS

Nomenclature

Enzyme Commission (EC) numbers used for enzymes are those listed by the International Union of Pure and Applied Chemistry and International Union of Biochemistry, published in *Enzyme Nomenclature* (1973 edition).

Names and abbreviations for nucleic acids, polynucleotides and their constituents are those recommended by the IUPAC and IUB Commission, published in the *Biochemical Journal,* **120,** 449–454 (1970).

Acceptor RNA   *see* TRANSFER RNA

Actin   Protein extracted from muscle. Mol. wt. about 70 000. In skeletal muscle molecules of actin form the 'thin filaments' of the myofibrils, associated with the 'thick filaments' of the major muscle protein myosin. G-actin is the spherical monomer whose repeated structure gives rise to the strand of F-actin. This actin filament has twin strands forming a right-handed double helix around a common axis. The protein tropomyosin lies in the grooves of the actin helix and with troponin and calcium ions forms a regulatory system for the actin–myosin interaction that is responsible for muscle contraction[82].

See also MYOSIN

Activating enzymes   *see* AMINOACYL-tRNA SYNTHETASES

Activation   Change induced in an amino acid molecule before its utilisation in cells for protein synthesis. By the action of activating enzyme in the presence of ATP, formation of the amino acid adenylate, which remains bound to the enzyme, takes place. Activation of an amino acid leads to its 'recognition' by a specific transfer RNA molecule, to which it is attached after trans-acetylation by the activating enzyme and transported to the ribosome for protein synthesis.

See also AMINOACYL-tRNA SYNTHETASES

Active site (Binding site; Catalytic site)   Region of a protein (enzyme) molecule at which binding occurs with the compound (substrate) whose structure is changed by the enzyme's activity. The groups composing the active site can be widely spaced on the polypeptide chain but are brought together by the molecular folding in the secondary and tertiary structures to give rise to a functional unit.

See also ALLOSTERIC TRANSITION

Actomyosin (Myosin B)   Protein forming the contractile component of muscle. It is a complex of the polypeptide chains of myosin and actin and possesses adenosine triphosphatase activity.

See also ACTIN; MYOSIN

Adaptor molecule   *see* TRANSFER RNA

Adenine (6-Aminopurine)   A derivative of the heterocyclic compound purine, forming a principal base in nucleic acids.

See also ADENOSINE; BASE

1

Adenosine　One of the ribonucleosides (containing the purine base adenine) making up nucleic acids and forming one unit of the genetic code (A). Adenosine 'pairs' non-covalently with the deoxyribonucleoside ribosylthymine (T) in deoxyribonucleic acid (DNA) and with uridine (U) in ribonucleic acid (RNA).

　　See also BASE

A-form (DNA-A)　Molecular conformation of DNA (DNA-A) at relative humidity less than 70%: x-ray analysis of the sodium salt of DNA-A shows a highly ordered crystalline form. At 90% relative humidity DNA-A becomes the paracrystalline B-form (DNA-B)[64].

　　See also B-FORM; C-FORM; DEOXYRIBONUCLEIC ACID

Allele　The changed form of a gene or locus occurring by mutation, in which deletions, substitutions or insertions have altered the original specific sequence of nucleotides.

　　See also LOCUS

Allosteric effector (Allosteric modulator)　A small molecule capable of interacting at a non-catalytic site (allosteric site, regulator or control site) on an enzyme molecule, with a consequent modulation of the enzyme's activity, or on other protein molecules, with a consequent change in the biological functioning of the protein.

Allosteric transition　A reversible molecular transition of a protein molecule that modifies the properties of the active site and changes the protein's biological activity.

Alpha helix　*see* α- HELIX

Amber triplet　Name given to the codon UAG after it was recognised that this terminal codon could be suppressed in the 'amber' mutant of *Escherichia coli* bacteriophage T4. (H. Bernstein was a discoverer of the T4 mutant selectively attacking a particular *E. coli* strain; *Bernstein*, German: amber.)

Ambiguous codon　A nucleotide triplet capable of coding for more than one amino acid ('miscoding'). Poly(U) (codon UUU) codes not only for phenylalanine but also to a restricted extent for leucine. Antiobiotics (streptomycin, kanamycin, neomycin) increase miscoding errors, as do variations in pH, temperature and ions in experimental systems.

　　See also CODON; GENETIC CODE

Amino acid sequence　Ordered arrangement of amino acids forming a specific polypeptide or protein molecule, determined by sequence analysis. This sequence expresses the primary structure of such a molecule. Insulin was the first protein to be analysed completely in this way[105].

Amino terminal (*N*-terminal residue)   Amino acid (residue) at one end of a polypeptide chain having its amino group free, i.e. not participating in peptide bond formation. Identification of amino group terminals in a protein undergoing controlled degradation is part of the procedure of end-group analysis by which amino acid sequences are determined.

Aminoacyl adenylate   *see* AMINOACYL-tRNA SYNTHETASES

Aminoacyl-tRNA synthetases   Each aminoacyl-tRNA synthetase (e.g. tyrosyl-tRNA synthetase; EC 6.1.1.1) 'activates' a specific amino acid by promoting its reaction with ATP to give an aminoacyl–adenylate–enzyme complex. The carboxyl group-activated amino acid is accepted by a specific transfer RNA (tRNA) before its inclusion in a polypeptide chain at the ribosome, the aminoacyl-tRNA synthetase also catalysing the transfer to the tRNA.

   See also ACTIVATION

Annealing   The slow cooling procedure that brings about the recombination of the separated strands of denatured ('melted') deoxyribonucleic acid from bacterial sources.

   See also DENATURATION; HYBRIDISATION TECHNIQUE

Anticodon   The triplet of bases in transfer RNA (tRNA) that is complementary to and will therefore pair with the codon in messenger RNA (mRNA). For example, in phenylalanine tRNA the anticodon is GAA (nucleosides guanosine and adenosine). The anticodon forms part of a 'loop' of nucleotides formed by non-helical regions of the RNA strand.

   See also CODON; GENETIC CODE; TRANSFER RNA; TRANSLATION

Anti-sense strand   The single DNA strand of the double helix of which the RNA formed by transcription is a copy. Its partner strand, complementary to the RNA, is the 'sense strand'.

   See also TRANSCRIPTION

Apoenzyme   The protein subunit of an enzyme molecule in those enzymes with molecules additionally containing a non-protein 'prosthetic group'.

Apurinic acid   The product of treatment of DNA with dilute mineral acid. The compound has the original DNA pyrimidine nucleotides but the purine nucleotides are replaced by deoxyribose phosphate units. Acid degradation of DNA has been used in determination of nucleotide sequences.

Bacteriophages (Phages) Simplest known viruses, with bacteria as hosts. Since in each 'phage' particle the single-stranded RNA genetic material specifies only three or four proteins the replication of phages has been investigated in detail. A complete, biologically active chromosome of the phage φX174 was synthesised enzymically *in vitro*, with the phage DNA used as a template, in 1967[34]. A complete gene sequence coding for a protein was first determined[80] in respect of the coliphage MS2 (host organism *Escherichia coli)* 'coat protein'.

　See also LYSOGENY; RIBONUCLEIC ACID BACTERIOPHAGES

Base Nitrogen-containing cyclic compound produced on hydrolysis of nucleic acid. Nucleic acid bases are derivatives of either pyrimidine (cytosine, uracil, thymine) or purine (adenine, guanine). *N*-Methyl- and *C*-methyl-substituted adenine and guanine are 'rare bases' found in small amounts in various species of RNA.

　See also PURINE BASES; PYRIMIDINE BASES

Base pairing Purine and pyrimidine bases of the nucleotides forming the two strands of a double helix are paired by hydrogen bonds between them around the helical axis. Specifically, adenine is paired with thymine and guanine with cytosine, for spatial reasons, requirements that also determine what the order of bases must be in a second (complementary) strand formed from the first strand. Theoretical considerations of base pairing in single-stranded RNA species of known base sequences have enabled the secondary structures of these molecules to be postulated.

　See also DOUBLE HELIX

Base sequence The specific order and identities of component nucleotides in a polynucleotide denoted by the purine and pyrimidine bases they contain and expressed by the single letter abbreviations used for the nucleosides. Many base sequences for small RNA molecules are known. Base sequences for DNA species have been determined by various techniques indirectly after (1) transcription of the DNA *in vitro* with RNA polymerase *(Escherichia coli)* and then analysing the transcribed RNA, or (2) synthesis of a labelled complementary DNA chain with DNA polymerase I *(E. coli)* and analysing this, or by the direct method used[100, 138] for phage φX174 DNA, involving sequential enzymic hydrolysis with endonuclease and exonuclease and analysis of the final products.

　See also BASE; BASE PAIRING

Base stacking Arrangement of the nucleotide purine and pyrimidine

4

bases in the Watson–Crick double-helix model. The bases are at the centre of the molecule, at right angles to the helical axis and stacked 0.34 nm apart. In the two strands the paired bases are coplanar, the purine base of one strand being hydrogen-bonded to a pyrimidine base of the other strand. Vertical forces (van der Waals and depolarisation) interacting between the stacked bases contribute to the stability of the helix.

See also DOUBLE HELIX

Beta configuration   *see* FIBROUS PROTEINS

Beta structure   *see* β- PLEATED SHEET

B-form (DNA-B)   In aqueous medium the molecular conformation of DNA is the paracrystalline form[64] known as DNA-B. At relative humidity less than 44% DNA-B changes to the C-form of DNA.

See also A-FORM; C-FORM; DEOXYRIBONUCLEIC ACID

Binding site   *see* ACTIVE SITE

'Breathing'   Transient   dynamic   changes   of   structure   in macromolecules, changes suggested to facilitate hydrogen exchange from the interior of the structure and being of possible biological significance in the function of the molecule[28, 94]. Opening and closing of the structure of DNA for a particular nucleotide pair could be a control system for access of enzymes to the molecule.

CAP factor   *see* CYCLIC AMP RECEPTOR PROTEIN

Capsid   Protein shell of a simple virus particle. The capsid protein surrounds the innermost 'core' of polynucleotide (RNA or DNA) and consists of units called capsomeres. It is enclosed by the outer membrane ('sheath') of the virus particle.

Carboxyl terminal (C-terminal residue)   Amino acid (residue) at one end of a polypeptide chain having its carboxyl group free, i.e. not participating in peptide bond formation. Identification of carboxyl-group terminals in a protein undergoing controlled degradation is part of the procedure of end-group analysis by which amino acid sequences are determined.

Catalytic site   *see* ACTIVE SITE

Catenated molecules   Cyclic ('circular') molecules joined in the manner of links in a chain. Interlinked cyclic DNA molecules occur in infected *Escherichia coli* cells when replication of the infecting

5

phage DNA is blocked by the use of a mutant host (*rec* A)[5].
Mitochondrial cyclic DNA molecules may also be catenated[137].

'Central dogma' Molecular explanation of genetics involving the hypothesis[134] that genetic information inside cells is represented in polynucleotide molecules by linear sequences of bases coding for amino acid sequences of protein molecules. This has been summarised as: DNA→RNA→Protein.

See also GENETIC CODE

C-form (DNA-C) Molecular conformation of DNA (DNA-C) at relative humidity less than 44%[77].

See also A-FORM; B-FORM; DEOXYRIBONUCLEIC ACID

Chain initiation see PROTEIN SYNTHESIS

Chloroplasts Disc-like bodies about 5 nm in diameter, containing chlorophyll, seen in the cytoplasm of cells of green plants. Photosynthesis of starch occurs within the chloroplast, even when isolated *in vitro*. Since polynucleotides occur in chloroplasts it is possible that a genetic independence from the cell nucleus exists.

Chromatid One folded fibre (one DNA–protein complex) of the identical pair formed at the metaphase stage of cell division from a single chromosome by replication. The two chromatids are attached at unreplicated regions before their separation as the two daughter chromosomes.

Chromatin (Nucleohistone) Complex of DNA and protein (deoxyribonucleoprotein) obtainable from cell nuclei. Histone is the major protein of this complex.

See also HISTONES

Chromosome Originally the name used to describe the discrete strongly staining units observed microscopically in the nuclei of animal and plant cells at cell division. These bodies are DNA–protein complexes (deoxyribonucleoprotein). Now also applied to the genetic material of prokaryotes, for example the DNA of *Escherichia coli* is in one strand, 1000 times the length of the bacterium but with a compact 'chromosome' formation apparently not involving protein. Models of the chromosome represent it as various arrangements of irregularly folded fibres of uneven diameter[20, 24].

Circular DNA (Cyclic DNA) Covalently closed DNA chains[43]. Both single-stranded (rare) and duplex forms are known[95]. Circular duplex DNA species occur in many viruses and mitochondrial DNA in all metazoan species examined is in this form [58, 98, 137]. Replication of the single circular DNA molecule (3×10^6 base pairs) forming the chromosome of *Escherichia coli* takes place in both directions from one fixed origin[7, 78].

See also SINGLE-STRANDED DNA; SUPERHELIX

Cistron   Shortest length of DNA carrying information adequate to define the composition of a single protein[6]. Nucleotides of the cistron constitute a functional unit for protein synthesis. A protein containing 500 amino acids would be represented by a cistron of 1500 nucleotides, i.e. 500 codons. Now used synonymously with 'gene'.

See also CODON; MESSENGER RNA; POLYCISTRONIC MESSENGER

Cloverleaf pattern   Description of the postulated two-dimensional outline of the transfer RNA (tRNA) molecule, formed by loops and folds of the incompletely base-paired polynucleotide chain[46].

See also TRANSFER RNA

Coat protein   A protein layer, enclosing the genetic material (DNA or RNA) making up the virus or bacteriophage particle. The protein 'recognises' the host cell, to which it adheres whilst the viral polynucleotide invades the cell. In some viruses one gene codes for the coat protein, thus providing relatively simple experimental material. For *Escherichia coli* bacteriophage R17 a sequence of 57 RNA nucleotide bases coding for a 19 amino acids length of the phage coat protein was isolated by F. Sanger's group in 1969[1]. The first complete gene sequence coding for a protein was determined from the RNA of coliphage MS2, in which 49 codons specify the phage coat polypeptide of 129 amino acids length[80].

Codegenerate codon   *see* DEGENERATE CODE

Codon   A group of three nucleotides (trinucleotide, trimer, triplet) in a specific sequence and selection, which codes for an amino acid or serves to signal some change in the translation process ('punctuation'). The trinucleotide, as part of the polynucleotide strand known as messenger ribonucleic acid (mRNA), determines the linear position of a particular amino acid in the chain of amino acids (polypeptide) assembled on the mRNA strand during synthesis of a protein. For example, the codon for the amino acid methionine is the triplet AUG (nucleosides adenosine, uridine and guanosine), and for tryptophan the codon is UGG. Other amino acids have more than one codon, the third base of each triplet usually being the one that is different (e.g. threonine codons are ACA, ACG, ACC and ACU; C is cytidine) and it is usually the first two bases of a codon that are characteristic of the amino acid.

See also AMBIGUOUS CODON; ANTICODON; GENETIC CODE; NONSENSE CODON; TRANSLATION

Cohesive terminus The ends of DNA single-stranded molecules that show mutual complementarity in nucleotide sequences, so that they may join by base pairing, circular molecules being formed. It has been postulated that end-to-end joining of partially separated duplex DNA molecules can occur in the same way.

Coiled coil *see* SUPERHELIX

Colicins Proteins secreted by certain strains of *Escherichia coli* that will kill other sensitive strains of the bacterium, by inactivation of ribosomes. The inactivation is by ribonuclease action on 16S RNA, this enzyme becoming operative when colicins bind to the surface of the bacterium.

Coliphage *see* BACTERIOPHAGES

Collagen Fibrous protein of high tensile strength in its organised state, estimated to form (as connective tissue including bone) about one-third of total mammalian body protein. Every third amino acid in most of the primary structure sequence is glycine, with proline and hydroxyproline making 20–25% of the molecule. α_1 and α_2 polypeptide chains are recognised (2:1 molar ratio) and there are 1050 amino acids in a chain. The protein is a triple helix with the three helical polypeptide chains forming a three-stranded 'rope'[96]. Aggregates of the molecules aligned in parallel and with homologous regions in register produce a regular cross-banding seen in collagen fibrils (20–100 nm diameter) on electron microscopy. On heating, collagen is denatured, the molecular structure changing to the random coil state. Synthetic polytripeptides have been prepared, such as poly(Gly-Pro-Pro)[128], which show x-ray diffraction patterns like those of collagen.

Complementary base sequence Sequence of nucleotides in a polynucleotide chain, as determined by the order of purine and pyrimidine bases of the nucleotides in the parent chain. Adenine pairs with thymine and guanine pairs with cytosine, and these pairs are complementary.

   See also BASE PAIRING

Complementary transcription An unusual process of transcription involving both strands of DNA shown to exist for phage lambda[115], in which one gene can yield RNA by transcription in opposite directions at different times.

   See also TRANSCRIPTION

Concatemer *see* CONCATENATE

Concatenate (Concatemer) Structure consisting of a number of genomes connected end to end.

   See also CATENATED MOLECULES

8

Conjugation (bacterial) ('Mating') Process involving contact between two bacterial cells with passage of deoxyribonucleic acid genetic material (fertility factor) from one (F^+) cell to the other (F^-) cell[54].

See also FERTILITY FACTOR

Copolymer A polymeric sequence of units (monomers, e.g. nucleotides) of two or more types, e.g. the synthesisable poly(A, C), poly(A, G, C, U). Copolymers may be random, of stated [e.g. poly(A_3, C_2)] or unstated proportions, or (for two nucleotide types) alternating [designated, e.g. poly(A-C)].

Corepressor A specific metabolite which by combination with a protein (apo-repressor) synthesised by a regulator gene can cause the protein to bind to the operator gene region of a DNA chain. This binding prevents the production of enzyme by the structural genes of the operon affected by the repressor.

See also OPERON; REPRESSOR

CRP (crp) factor *see* CYCLIC AMP RECEPTOR PROTEIN

C-Terminal *see* CARBOXYL TERMINAL

'Cut' A double-stranded break in a duplex polynucleotide, in distinction to the single-strand scission or 'nick'.

Cyclic AMP (Adenosine 3′ : 5′-cyclic monophosphate) Monophosphate of the ribonucleoside adenosine, in which the atoms of the phosphate group are arranged in a ring, for which a number of intracellular regulatory functions have been postulated. These affect cellular secretion processes, enzymic storage of carbohydrates and fats and cell division. Cyclic AMP stimulates expression of genes in *Escherichia coli*, bringing about transcription. In cells transformed by cancer-causing viruses in tissue culture normal growth has been shown to be restored by the presence of cyclic AMP.

Cyclic AMP receptor protein [Catabolite gene activation protein (CAP factor); CRP factor] Protein present in *Escherichia coli* demonstrated to be necessary to bind cyclic AMP and initiate transcription. There appears to be many factors of this type in bacteria, acting as controllers of gene expression (psi factors).

See also CYCLIC AMP

Cyclic coil *see* RANDOM COIL

Cyclic DNA *see* CIRCULAR DNA

Cytidine One of the ribonucleosides (containing the pyrimidine base cytosine) making up nucleic acids and forming one unit of the genetic code (C). Cytidine 'pairs' non-covalently with guanosine (G) in deoxyribonucleic acid and ribonucleic acid.

See also BASE

Cytosine (2-Hydroxy-6-aminopyrimidine) A derivative of the heterocyclic compound pyrimidine, forming a principal base in nucleic acids (principally ribonucleic acid).

See also BASE; CYTIDINE

Degenerate code As the nucleotide triplets (64 possible arrangements) coding for amino acids (20) are not completely specific for individual amino acids the genetic code is 'degenerate'. For example, the six leucine codons are CUA, CUG, CUC, CUU, UUA, UUG. Any two codons specifying the same amino acid are 'synonymous' or 'codegenerate' codons.

See also CODON; GENETIC CODE; NONSENSE CODON; WOBBLE HYPOTHESIS

Deletion The loss of a base pair from the double-stranded DNA helix, responsible for errors in transcription. Induced deletions have been used to check the genetic code theory, as has study of naturally occurring deletions that produce mutations represented by the 'molecular diseases'.

See also INSERTION; REPLACEMENT; TRANSCRIPTION

Denaturation ('Melting') A change in the naturally occurring ('native') structure of biological macromolecules. For example, separation of double-stranded DNA into single strands by heating in solution to above the melting temperature (T_m); disruption of the α-helix (helix–coil transition) produces the single-stranded 'random coils'[75, 76].

See also ANNEALING; HYBRIDISATION TECHNIQUE

Deoxyribonuclease (DNase; DNAase; Streptodornase) Enzyme that degrades DNA, isolated in pure form from many types of cells. It is an endonuclease, with the phosphodiester bond as the site of action in the DNA molecule. Small groups of nucleotides are the end-products, deoxyribonuclease I types yielding 5′-phosphoryl-terminated oligonucleotides and deoxyribonuclease II types yielding 3′-phosphoryl-terminated oligonucleotides. Deoxyribonuclease I (deoxyribonucleate oligonucleotidohydrolase: EC 3.1.4.5) from the pancreas requires magnesium ions and has optimum pH 6.8–9.2. Deoxyribonuclease II (deoxyribonucleate 3′-nucleotidohydrolase: EC 3.1.4.6) from thymus or spleen is inhibited by magnesium ions and has optimum pH 4.5–5.5. Deoxyribonuclease is responsible for excision of abnormal sections of DNA in the DNA repair mechanism.

Deoxyribonucleic acid (DNA) A biological polymer of deoxyribonucleotides, of high molecular weight (10^6–10^9), associated with the protein (deoxyribonucleoprotein) of chromosomes. In animal and plant cell nuclei it is the main repository of genetic information and is also recognised as the genetic material of bacteria and viruses. Extranuclear DNA is found in mitochondria and chloroplasts. The smallest known natural species of DNA (genomes of bacteriophages φX and fd) are more than 5000 residues in length. DNA occurs as unbranched strands, single or double, and in a cyclic ('circular') form, single or double. Molar proportions of the four nucleotide bases of DNA vary with species. Apparently universal features of the base composition of double-stranded DNA are[15]: total purines equal the total number of pyrimidines; adenine and thymine are equimolar (A = T); guanine and cytosine are equimolar (G = C); total amino-containing bases (A, C) equal in number the total oxo-containing bases (G, T). The double-stranded (duplex) DNA molecule forms a double helix in which the strands are 'anti-parallel'. The strands are complementary as a result of the base pairing shown (A to T, G to C) and when separated (unwound) each can become the template for replication of a new complementary strand. Two duplexes are then formed from the original duplex and are identical with it: each new fibre pair has one mother strand and one newly synthesised strand, the base sequence of the latter having been determined by the sequence of complementary bases on the parent strand[79] ('semiconservative replication'). The secondary structure of DNA alters according to the water content of the sample. Native DNA in aqueous medium is in the paracrystalline B-form (DNA-B). Synthesis of DNA is known to be a discontinuous replication process[86], involving the formation of short-chain molecules that are subsequently enzymically joined together by a ligase. There is evidence that in *Escherichia coli* each DNA segment requires an RNA sequence to initiate its synthesis[45], an RNA chain of 50–100 nucleotides being present initially at the C-5′ end of the DNA segment and subsequently removed[118].

cDNA: transcribed (complementary) DNA obtained enzymically with reverse transcriptase from an RNA virus.

O-DNA: oxidised DNA, in which the bases are in the ureido form.

    See also A-FORM; C-FORM; CHROMOSOME; CYCLIC DNA; DEOXYRIBONUCLEOPROTEIN; DOUBLE HELIX; OKAZAKI FRAGMENTS

11

Deoxyribonucleic acid ligase (DNA ligase; DNA joinase; Polynucleotide synthetase, EC 6.5.1.1 and 2) Enzyme(s) from *Escherichia coli* capable of covalently joining short strands of DNA ('Okazaki fragments') formed by the polymerising activity of DNA polymerase, thus completing the synthesis of the polynucleotide. The enzyme(s) also repair single-strand breaks in a DNA chain and are adenosine triphosphate- (EC 6.5.1.1) or diphosphopyridine nucleotide- (EC 6.5.1.2) dependent[32]. Mammalian and plant cells[4, 42] have been shown to possess DNA ligase (adenosine triphosphate-dependent) with the same replicative function.

Deoxyribonucleic acid polymerase (DNA nucleotidyltransferase; Deoxynucleoside triphosphate–DNA deoxynucleotidyltransferase, EC 2.7.7.7; Duplicase; Kornberg enzyme) Enzyme, first purified from extracts of *Escherichia coli*, capable of catalysing the replication of DNA and repair of damage to DNA molecules. The enzyme binds at 'nicks' in the DNA strand. Its single polypeptide chain (mol. wt. 109000) yields two fragments (mol. wt. 76000 and 36000) on limited proteolysis[61]. These yield DNA polymerases I, II and III. Two activities are shown by the larger fragment[13, 107, 108]: extension of a DNA chain 'primer' by polymerisation and degradation of DNA to nucleotides from the C-3′ hydroxyl to the C-5′ phosphate end of the chain (3′→5′ exonuclease activity). The smaller fragment degrades only double-stranded DNA, from the C-5′ end to the C-3′ end, producing nucleoside 5′-phosphates (5′→3′ exonuclease activity). DNA polymerase from avian myeloblastosis virus has been shown to be capable of copying globin messenger RNA of rabbit[104, 130] and human[56]. Mammalian cells have been shown to possess at least two DNA polymerases responsible for polymerisation of the DNA monomers.

Deoxyribonucleic acid (DNA) viruses *see* VIRUS

Deoxyribonucleoprotein Protein held by electrostatic forces in association with deoxyribonucleic acid (DNA) of the nucleus. The low-molecular-weight protamines (nucleoprotamines) and histones (nucleohistones), both basic proteins, are isolated from cell nuclei and sperm-heads. Nucleoprotamines are thought to occupy the minor (shallow) groove of the DNA helix. Part at least of the nucleohistones occupy the major (deep) groove. Nucleohistones have been proposed to be repressors to the expression of certain tracts of the DNA in gene translation. Non-histone proteins of

12

deoxyribonucleoprotein are known to control the binding of steroid hormone–receptor complexes to the DNA[73].

See also DOUBLE HELIX; GROOVE; HISTONES; PROTAMINES

Deoxyribonucleoside and deoxyribonucleotide see NUCLEOSIDE AND NUCLEOTIDE

Deoxyribose Pentose sugar derivative present in all cells as a component of deoxyribonucleic acid in polyribonucleotides. Deoxyribose phosphate molecules form the chain bonded by phosphodiester links making the backbone of polyribonucleotide fibres as in DNA. Deoxyribose is formed from ribose after this has been incorporated into a nucleotide (ribose–base–phosphate).

Dimer A molecule resulting from the association of two similar relatively simpler molecules. (1) Dimer formation can occur between adjacent pyrimidine bases on one strand of DNA, particularly thymine dimers, as the result of ultraviolet irradiation of bacterial cells, and prevents normal DNA replication and reproduction of the organism. (2) A protein dimer consists of two identical polypeptide subunits (protomers or monomers).

Diploidy Cell state in which the chromosomes in the nucleus are paired as homologous partners. The number of chromosomes is thus twice that of the haploid cell.

See also HAPLOIDY

DNA, cDNA and terms prefixed by DNA see DEOXYRIBONUCLEIC ACID

DNA-dependent RNA polymerase see RIBONUCLEIC ACID (RNA) POLYMERASE

DNA–RNA hybrid see HYBRIDISATION TECHNIQUE

Double helical double helix Postulated structure for globular DNA in which a left-handed helix is formed from a loop of double-stranded DNA[20].

Double helix Model of the molecular arrangement of double-stranded DNA first proposed by Watson and Crick[134]. Each of the two polynucleotide chains is a right-handed helix; the pair are separable only by unwinding. Base pairing by hydrogen bonds (adenine to thymine, guanine to cytosine) keeps the chains together, and the base-stacking forces between adjacent bases of the nucleotides within each chain give the helix stability. Bases of the constituent nucleotides project towards the axis of the helix. The chains are 'anti-parallel', i.e. the C-5' end of one chain is base paired with the C-3' end of the other (complementary) chain. The

13

'pitch' of the helix is 3.4 nm, ten bases occupying this stretch of strand, each stacked 0.34 nm apart. A crystal of the dinucleotide adenosyl 3',5'-uridine phosphate has been revealed in atomic detail by x-ray analysis to form a right-handed, anti-parallel double helix with adenine and uracil held together by hydrogen bonds as required by the Watson–Crick model[103].

See also BASE STACKING; DEOXYRIBONUCLEIC ACID

Drift region   Region at the beginning of a transcriptional unit, before the site of activity of RNA polymerase, which can be occupied by the transcribing enzyme molecules without production of RNA chains[9].

Duplex   A double strand of polynucleotides, as in the double helix of deoxyribonucleic acid (DNA) or in the short looped sections of single-stranded ribonucleic acid (RNA) species in which base pairing is present.

See also BASE PAIRING; DOUBLE HELIX

Effectors   Metabolites which specifically affect the production of proteins (enzymes) in a bacterial cell. Two types are recognised: inducers (inactivate repressors of protein synthesis) and co-repressors (activate repressors of protein synthesis).

See also COREPRESSOR; INDUCTION; OPERON; REPRESSOR

Elongation factors   Cellular agents concerned with growth in length of macromolecule chains. (1) Proteins capable of controlling DNA synthesis, suggested to be the histones of deoxyribonucleo-protein[135]. (2) Proteins isolated from bacteriophages and bacteria and shown to be required for the formation of peptide bonds in polypeptide and protein synthesis (G, S_{1-3} and T_s, T_u factors).

Endonucleases   Enzymes (phosphodiesterases) capable of hydrolysing polynucleotides, specific types with ribonucleic acid (RNA) or deoxyribonucleic acid (DNA) as substrate being recognised.

See also DEOXYRIBONUCLEASE; RIBONUCLEASE

Entatic state   Condition of activation occurring in one region of a protein molecule as a result of conformational requirements for stability of other parts of the molecule ('geometrical and electronic strain')[129].

14

Episome   Genetic (deoxyribonucleic acid) element in bacterial cells that is additional to the cell genome and is capable of independent replication[53]. The episome can be attached to the chromosome and replicate with it. Episomes have been suspected to occur in eukaryotic cells.

See also PLASMID

Euchromatin   The dispersed, less dense form of chromatin (DNA–protein) seen in the cell nucleus at interphase, containing the 'euchromatic DNA', presumed to represent the active region of the chromosome material.

See also HETEROCHROMATIN

Eukaryote (Eukaryotic cell; Eukaryocyte)   An organism with cells that have a discrete membrane to the nucleus, enclosing the genetic material[117]. Cells other than those of bacteria and blue–green algae are of this type.

See also PROKARYOTE

Exonucleases   Enzymes (phosphodiesterases) capable of hydrolysing polynucleotides to yield mononucleotides, the degradation of the chain occurring sequentially from one end to the other, either in the C-5′ to the C-3′ direction (5′→3′ exonuclease) or in the reverse direction (3′→5′ exonuclease).

See also DEOXYRIBONUCLEIC ACID POLYMERASE

Fertility factor (F-factor; Sex factor)   Genetic element (plasmid) of circular deoxyribonucleic acid occurring within *Escherichia coli* K-12 cells that is capable of the passage of replicative material in the process of conjugation ('mating') to another cell[54]. The state of harbouring the F-factor is denoted as F+; F− is the state of a cell without the F-factor, and such a cell (loosely, 'female bacterium') can be infected with the F-factor from a donor F+ cell (loosely, 'male bacterium').

F-factor   *see* FERTILITY FACTOR

Fibrous DNA   Stretches of relatively uncoiled double strands of DNA suggested to be regions of specific base sequences of coding rather than gene control[20].

See also GLOBULAR DNA

Fibrous proteins   Proteins with elongated thread-like polypeptide chains forming their molecules, e.g. silk fibroin, hair and wool keratin, myosin, epidermin and fibrinogen.

15

'Flower' model   Name given to the elucidated complete secondary structure model of the coat protein cistron sequence for RNA of coliphage MS2[80], from the appearance of the double-stranded 'stem' terminating in a set of radiating double-stranded 'petal' loops.

Folding (molecular)   Essentially a thermodynamic process dictated by the necessity of attaining the most stable structure in a protein molecule, which is determined by the molecule's amino acid sequence (primary structure).

N-Formylmethionine   This formylated derivative of the amino acid methionine in conjunction with methionine transfer RNA (tRNA^{Met}) is known to be the initiator of polypeptide synthesis in *Escherichia coli*[66]. Placement of the N-formylmethionyl-tRNA is determined by the codons AUG or GUG at the beginning of the RNA sequence (AUG or GUG within the length of the sequence attracts non-formylated methionyl-tRNA). After initiation, the formylated compound is detached from the amino end of the polypeptide chain.

Frameshift mutation (Phase-shift mutation)   Change resulting from addition or deletion of nucleotides in numbers other than three (triplets; codons), which moves the translation 'reading frame' so that a new set of codons beyond the point of abnormality in the messenger RNA is 'read'[21]. The insertion of an incorrect amino acid sequence in the forming polypeptide chain is a result of this. Compensation for the mutation can occur by the phenomenon of suppression.

   See also SUPPRESSION; TRANSLATION

Gene   Conventional unit of hereditary material responsible for the passage of parental characters of offspring, recognisable by the phenomenon of spontaneous change (mutation) to an alternative form. Now used interchangeably with the term cistron to indicate the unit of genetic function; for example, a sequence of nucleotides of DNA adequate to define a protein's amino acid composition in its entirety.

   See also CISTRON; OPERON

Genetic code   Represented by nucleotides having four different types of base and used in DNA for storage of information about the amino acid sequences of a specific protein. There are 20 different amino acids, each coded by a sequence of three nucleotide bases (the codon). This arrangement gives $4^3 = 64$ possible combinations, and more than one triplet may code for a particular amino acid[21]. Three of the 64 possible codons, UAA, UAG and UGA, are known as 'nonsense codons' as they do not represent amino acids and serve as 'chain-terminating signals'. The complete code is shown in *Table 1*.

See also ANTICODON; CODON; DEOXYRIBONUCLEIC ACID; NUCLEOSIDE AND NUCLEOTIDE

Table 1 THE GENETIC CODE

Each codon for an amino acid is a nucleotide triplet: the letter (U, C, A or G) for the first nucleotide base is read in the left-hand column, that for the second nucleotide base is in the top line and that for the third nucleotide base is in the right-hand column. Amino acids: Ala, alanine; Arg, arginine; Asn, asparagine; Cys, cysteine; Gln, glutamine; Glu, glutamic acid; Gly, glycine; His, histidine; Ile, isoleucine; Leu, leucine; Lys, lysine; Met, methionine; Phe, phenylalanine; Ser, serine; Thr, threonine; Trp, tryptophan; Tyr, tyrosine; Val, valine.

Second	U	C	A	G	
First					*Third*
	Phe	Ser	Tyr	Cys	U
	Phe	Ser	Tyr	Cys	C
U	Leu	Ser	Nonsense (ochre)	Nonsense	A
	Leu	Ser	Nonsense (amber)	Trp	G
	Leu	Pro	His	Arg	U
C	Leu	Pro	His	Arg	C
	Leu	Pro	Gln	Arg	A
	Leu	Pro	Gln	Arg	G
	Ile	Thr	Asn	Ser	U
A	Ile	Thr	Asn	Ser	C
	Ile	Thr	Lys	Arg	A
	Met	Thr	Lys	Arg	G
	Val	Ala	Asp	Gly	U
G	Val	Ala	Asp	Gly	C
	Val	Ala	Glu	Gly	A
	Val	Ala	Glu	Gly	G

Genome   The complete set of genes that is replicated within a cell.

Genotype   Genetic content of an organism, i.e. the genes possessed, which might differ from characteristics or properties (phenotype) that it exhibits.

G-factor see TRANSLOCASE

Giant RNA (Pre-mRNA; HnRNA; D-RNA; dRNA; mIRNA) A nuclear 'DNA-like' RNA of very high molecular weight representing the precursor of messenger RNA (mRNA) or pre-mRNA.

Globin see HAEMOGLOBIN; MYOGLOBIN

Globular DNA Compact arrangement of duplex DNA formed by additional twisting of the fibre (supercoiling) into a superhelix or double helical double helix. It has been suggested[20] that such structures, controllers in function, occur in the band region of chromatids and are continuous with fibrous DNA of the interbands which represent the stretches used for coding.

Globular proteins (corpuscular proteins) Proteins whose molecules are revealed by x-ray studies to have a pseudo-spherical shape arising from the complex folding (tertiary structure) of their polypeptide chains. Plasma proteins, haemoglobin, myoglobin and enzymes are all globular proteins.

α-Globulin see HAPTOGLOBIN

Glycoprotein Proteins in molecular association with carbohydrate (hexosamine sugar). Phytagglutinins (lectins) are examples of interest to cell biologists because of the property of binding to the surface of transformed cells, and they give a means of identifying and further experimenting with such cells. Concanavalin A is a lectin glycoprotein used in this way.

Groove In the double helix of deoxyribonucleic acid two grooves are recognised, the major, deep groove of the coiling double chain and the minor, shallow groove separating the pair of chains and marking the position of the base-paired nucleotides. These grooves are considered to be the positions occupied by the nuclear proteins.
  See also DOUBLE HELIX; DEOXYRIBONUCLEOPROTEIN

Guanine (2-Aminohypoxanthine) A derivative of the heterocyclic compound purine, forming a principal base in nucleic acids.
  See also BASE; GUANOSINE

Guanosine One of the ribonucleosides (containing the purine base guanine) making up nucleic acids and forming one unit of the genetic code (G). Guanosine 'pairs' non-covalently with cytidine (C) in deoxyribonucleic acid and ribonucleic acid.
  See also BASE

Haemoglobin Protein with the prosthetic group haem, having the property of reversibly combining with oxygen and, to a lesser extent, carbon dioxide, occurring predominantly in vertebrate blood (red blood cells, erythrocytes). The molecule is approximately spherical with four sub-units, two α-chains and two β-chains, each having over 140 amino acids forming irregularly coiled helices and each surrounding a haem molecule characterised by its ferrous iron[92]. Approximate mol. wt. 65 000. Combination with oxygen causes changes in radius of the iron atoms with consequent rearrangement of the haem units and globin chains (allosteric mechanism)[91]. The complete amino acid sequence of a haemoglobin has been analysed[10]. Studies of sickle cell anaemia revealed that a gene defect causing a replacement of glutamic acid by valine in one of the β-chains of haemoglobin is responsible for the disease, and a large number of rarer disease-producing abnormalities in haemo-globin are known.

See also MOLECULAR DISEASES

Haploidy The cell state in which only a single, unpaired, set of chromosomes is present in the nucleus. This state is produced by 'reduction division' (meiosis) of the diploid cell to form gametes (e.g. sperms, ova).

Hapten A small molecule capable of acting as an antigen (provoking an antibody response) in the animal body when the substance is linked to a macromolecule.

Haptoglobin (α-Globulin) Protein of human plasma capable of binding free and haemoglobin. The molecule consists of two pairs of non-identical polypeptide chains, bonded by disulphide links.[113]

α-Helix Regular coil configuration of the polypeptide chains of fibrous proteins (e.g. keratin, epidermin, myosin, fibrinogen). W. T. Astbury described[3] the characteristic x-ray diffraction pattern of fibrous proteins as the α-pattern, and L. Pauling showed [88, 89], first for α-keratin, that this pattern arises from the helical secondary molecular structure, which he named the α-helix. Each amide group forms hydrogen bonds with the amide groups removed by three from it in each direction along the polypeptide chain and these bonds hold the structure rigid.

See also DOUBLE HELIX; FIBROUS PROTEINS

Helix–coil transition Change occurring in regularly coiled molecules (helix conformations) that produces the irregularly coiled fibre ('random coil'). This change is seen in the double-stranded DNA helix when it is heated in solution to above the 'melting temperature' (T_m); the strands separate (unwind) and the regular coil of the helix becomes the 'random coil'. Denaturation of some proteins is held to involve such a change and a partial helix—coil transition has been invoked for myosin in a proposed mechanism of muscular contraction[41].

See also DENATURATION

Heterochromatin ('Dense chromatin') The clumped type of chromatin (DNA—protein) in the nucleus at interphase, described as dense chromatin and composed of 'heterochromatic DNA'. It is differentiated from the other type of chromatin (euchromatin) by staining techniques and its DNA is known to be replicated at a later stage than the euchromatin DNA before cell division. Associated with it is 'satellite DNA'. The function of this region is unknown, although it has been suggested that it represents the specifically inactivated ('switched off') part of the genome. The sex chromosomes of animals appear to be almost entirely heterochromatin.

See also EUCHROMATIN

Heteroduplex A double-stranded (duplex) deoxyribonucleic acid (DNA) molecule in which, unlike the usual arrangement, the two strands show sections that differ in the information represented by a cistron of each (i.e. alleles of the cistron). This model was proposed[67] for the chromosome of crosses between bacteriophages of *Escherichia coli*.

Heterogeneous RNA (Heterogeneous nuclear RNA; HnRNA; H-RNA) Large molecules of ribonucleic acid (RNA) formed in the nuclei of metabolically active cells and believed to be the precursor form of messenger RNA (mRNA).

Heterokaryon Hybrid cell formed artificially in tissue culture by introduction of a nucleus from a somatic cell of one species into the cell of another species. For example, avian erythrocyte nucleus into a human Hela tumour cell.

Histones Basic proteins occurring in complex with DNA (nucleohistone) in cell nuclei, forming approximately half of the total mammalian chromosomal protein. Characterised by a high proportion of lysine and arginine and absence of tryptophan. Several classes of histones based on the relative abundance of lysine or arginine are recognised. One arginine-rich class has been shown to have an identical sequence of 102 amino acids in material from a wide variety of plants and animals, possibly indicating a function of fundamental biological significance[22, 23]. Functions suggested for histones in deoxyribonucleoprotein complexes include stabilisation of the tertiary structure of DNA, gene repression by limiting the regions of DNA used as template for messenger RNA formation, and action as chain-elongation factors in control of DNA synthesis[135].

HnRNA   *see* HETEROGENEOUS RNA

Holoenzyme   The complete enzyme molecule in those enzymes with molecules containing a non-protein 'prosthetic group' as well as the protein 'apoenzyme'.

Homopolymer   A polymeric sequence of units (monomers, e.g. a nucleotide) of the same type, e.g. polyuridylic acid [poly(U)]. Homopolymers of nucleotides can be synthesised *in vitro* by the action of DNA-dependent RNA polymerase in the presence of DNA, Mn^{2+} and the base phosphate.

H-RNA   *see* HETEROGENEOUS RNA

Hybridisation technique (Molecular hybridisation)   A means of testing the identity of base sequences in two polynucleotide chains from different sources. Two chains of DNA having complementary base sequences will form a double-stranded structure on being cooled together in solution, and so will a DNA chain and a RNA chain that are complementary (forming a DNA–RNA hybrid)[39]. 'Melting' DNA, i.e. denaturation or separation of the double-stranded helix into single strands, occurs in solution when its temperature is raised above T_m (the melting temperature); 'annealing', i.e. recombination, occurs on slow cooling. Such hybridisation experiments with tumour virus polynucleotide and DNA from uninfected cells have been performed to determine whether it is possible for the DNA to code for the virus genes (formation of a stable hybrid is assumed to indicate that coding is possible).

Hyperchromicity   An increase in ultraviolet light absorption (E_{260}) by solutions of polynucleotides as a consequence of loss of the ordered secondary structure. This occurs when denaturation (helix–coil transition) of the double helix of DNA follows heating in solution to the 'melting temperature' (T_m).

Hypochromicity   A decrease in ultraviolet light absorption (E_{260}) accompanying changes within solutions of polynucleotides. This occurs with the formation of the ordered secondary structure (double helix) of DNA from a mixture of the two single polynucleotide strands.

Immunoglobulins (Ig) Group of high-molecular-weight proteins $(0.15-1 \times 10^6)$ classed as β- and (mainly) γ-globulins (by electrophoresis) that occur in vertebrate blood plasma and which represent antibodies. The γ-globulin molecules consist of 'heavy' and 'light' polypeptide chains, with a basic conformation (the 7S unit) of four chains, two 'heavy' and two 'light' (IgA, IgE, IgG, IgD). Some immunoglobulins have multiples of four chains: IgA (dimer) and IgM (pentamer).

Induction The starting or enhancement of synthesis of an enzyme by a cell, taking place upon the provision of the substrate for the enzyme (i.e. the 'inducer'). Depletion of substrate leads to cessation of enzyme synthesis ('de-induction'). 'Superinduction'[126] has been observed as an effect of the inhibitor of RNA synthesis, actinomycin D, which stops de-induction and the inductive action of steroid hormones on specific proteins[127].

See also OPERON

Informational macromolecules High-molecular-weight compounds (particularly DNA and RNA) present in all cells and representing the store of genetic instructions.

Informofers Special globular protein particles that form complexes with giant dRNA (DNA-like RNA) to give specific ribonucleoprotein of cell nuclei. A suggested function of informofers is facilitation of the action of processing enzymes (nucleases) on the dRNA that has been detached from the chromosomal template and is held on the surfaces of these proteins[70].

Informosomes Cell particles suggested to be complexes of messenger ribonucleic acid (mRNA) with protein (ribonucleoprotein) formed in the cytoplasm after production of mRNA in the nucleus[116].

Initiation factors Cellular components required for recognition of the (initiation) site on a template molecule for commencement of the process of transcription or protein synthesis. Specific initiation factors in *Escherichia coli* are known to bind the initiator transfer RNA (tRNA) for protein synthesis to ribosomes[66, 69, 93].

See also SIGMA FACTOR

Initiator codons Triplet nucleotides (codons) acting as 'start' signals for synthesis of polypeptide (from the amino end). For bacteria it has been shown[17, 74] that AUG at the beginning of the RNA sequence directs placement of formylated methionine by transfer RNA (although when within the sequence AUG codes for non-formylated methionine). Formylated methionine acts as polypeptide chain initiator in bacteria and is subsequently removed from the beginning leaving usually alanine or serine as the amino acid in the first position.

See also CODON

Insertion The addition of an extranumerary base pair to the double-stranded DNA helix, responsible for errors in transcription. Insertion of extra base pairs is a method of inducing mutations.

See also DELETION; REPLACEMENT; TRANSCRIPTION

Insulin Protein hormone, produced by β-cells of the islet tissue of the pancreas, governing metabolism of glucose. Bovine insulin was the first protein molecule to be completely defined structurally[105]. Its 51 amino acids occur in two chains (one of 21 and one of 30 amino acids) joined by sulphur (S–S) links. Mol. wt. about 6000. The rhombohedral insulin crystal is a hexamer, arranged as three dimers.

Isoacceptor One of two or more transfer RNA (tRNA) species capable of accepting the same amino acid (distinguished in abbreviated systems, as for example with the isoacceptors for alanine: $tRNA^{Ala}_1$, $tRNA^{Ala}_2$).

Jacob–Monod model *see* OPERON

Kappa factor A protein with two identical subunits that occurs in *Escherichia coli* and which blocks the transcription of bacteriophage T4, T5 and T7 DNA by the organism's RNA polymerase[106]. Kappa is thought to be a 'termination factor'.

Keratin Fibrous animal protein forming hair, wool, etc. Characterised by a high content of the sulphur-containing amino acid cystine (total sulphur content of the protein 3–6%) with disulphide cross-links accounting for the physically durable features of the protein. By x-ray diffraction analysis, fibres of the natural form (α-keratin) reveal the α-helix structure and in an isolated subunit it has been shown to be coiled to give the 'coiled coil' superhelix conformation[119]. Stretching of α-keratin gives the 'pleated sheet' form β-keratin[89].

Kornberg enzyme *see* DEOXYRIBONUCLEIC ACID (DNA) POLYMERASE

Lectins *see* PHYTAGGLUTININS

Ligase *see* POLYNUCLEOTIDE LIGASE

Light chain Polypeptide subunit(s) of a protein molecule having a lower molecular weight than another subunit(s) of the structure. For example, in immunoglobulin G two light chains (25 000 mol. wt.) and two heavy chains (55 000 mol. wt.) are present; light and heavy chains are also recognised in myosin.

Locus A specific nucleotide sequence in a polynucleotide determining the sequence of nucleotides in a specific ribonucleic acid molecule, or determining the sequence of amino acids in a specific polypeptide, or representing a regulatory or 'punctuation' function in translation. (This term is interchangeable with 'gene'.)

Lysogeny State created in bacterial cells infected by a 'temperate bacteriophage' without immediate lysis. The cell has the bacteriophage deoxyribonucleic acid (phage DNA: prophage) as part of its genome and this is transmitted to subsequent generations in which the prophage can become virulent and cause lysis of the cell with release of new bacteriophage. Repressor substances prevent expression of the prophage genes in the lysogenic cell.

See also BACTERIOPHAGES; REPRESSION

Lysosomes Cell particles of 200–600 nm diameter, obtainable by cell fractionation techniques, containing aggregations of enzymes.

Macromolecule A biological polymer of molecular weight in the thousands and in some exceeding a million. Proteins, polysaccharides and nucleic acids are examples of cellular macromolecules.

Meiosis   Process of 'reduction division' in nuclei of sex cells of higher organisms before formation of gametes. From the original diploid ($2n$ chromosomes) nuclear material of one cell, four gamete nuclei with the haploid (n chromosomes) number are produced. Two consecutive nuclear divisions are involved in meiosis; during the first of these, in which chromatids are formed, the phenomenon of 'cross-over', with exchange of segments between the chromatids ('genetic recombination'), can occur. In bacteria, protozoa and fungi undergoing conjugation, it is the newly formed cell that is diploid, so that meiosis follows the event to restore the haploid state of progeny cells.

See also MITOSIS; NUCLEUS

'Melting', 'Melting' temperature   *see* DENATURATION

Messenger RNA (mRNA) (D-RNA; Informational RNA; Complementary RNA; Transcript RNA; Translational RNA)   The minority form of RNA in cells (<5% of total RNA), with a molecular weight of at least 0.5×10^6. mRNA has a brief existence after its formation (transcription) as a complementary strand from a particular stretch of nuclear DNA (gene) by the action of the enzyme DNA-dependent RNA polymerase[36, 44, 81, 132]. mRNA in association with protein migrates to the ribosomes, where it acts as a template for the synthesis of a specific protein, i.e. mRNA carries amino acid sequence information, which is read from the C-5' end to the C-3' end of its chain. Groups of ribosomes along the thread of mRNA constitute the polyribosomes[57]. Difficulties of purification of mRNA have hindered base-sequence determinations for these ribonucleotides. Bacteriophages R17 and Qβ, in which RNA is both gene and messenger, have been used in sequencing studies[1].

See also   GIANT RNA; RIBONUCLEIC ACID; RIBOSOMES; TRANSCRIPTION

Methionine transfer RNA (tRNA^{Met})   Transfer RNA (tRNA) capable of accepting the amino acid methionine. Aminoacylation by enzymic action gives the compound methionyl-tRNA (Met-tRNA or Met-tRNA^{Met}). One type (tRNA$^{Met}_f$) can be formylated to give N-formylmethionyl-tRNA (fMet-tRNA$^{Met}_f$ or Met-tRNA^{fMet}). This form has been shown to initiate synthesis of protein polypeptide in *Escherichia coli* and other prokaryotes[66]. Met-tRNA$^{Met}_f$ is considered to be the universal initiator of translation in eukaryotes.

See also TRANSFER RNA

25

Micelle   Spherical or cylindrical aggregates of polar lipids formed in aqueous systems with a low lipid to water ratio.

Microsomes   Ultramicroscopic particles (16–150 nm diameter) of cytoplasmic origin obtainable on fractionation of cell homegenates in 0.25 M-sucrose by centrifugation at 100 000 g after removal of the nuclear and mitochondrial fractions. Electron-microscope examination of the microsomal fraction[87] shows that the particles are predominantly structures from the subcellular endoplasmic reticulum. The major amount of cytoplasmic ribonucleic acid occurs in the microsomal fraction and is separable from microsomes with protein as smaller isolated particles (ribosomes).

Minus strand   Complementary strand of polynucleotide formed by transcription from a specific polynucleotide ('plus') strand. In RNA bacteriophages the minus strand is formed from the 'parental' RNA template ('plus strand'), which, with the plus strand, produces the double-stranded (double-helix) RNA occurring in these bacteriophages.

Mitochondria (Chondriosomes)   Rod-shaped bodies (0.5–5.0 μm by 0.3–0.7 μm) visible by light-microscopy in the cytoplasm of cells (except bacteria and blue–green algae). Electron microscopy reveals mitochondrial membranes and internal septa (cristae) for each mitochondrion[102]. Mitochondria can be obtained by centrifuging cell homogenates in 0.25 M-sucrose at 8500 g after the nuclei have been removed. As the site of oxidative phosphorylation, the mitochondrion contains the multi-enzyme systems for the release and storage of energy from respiratory processes, and both deoxyribonucleic acid (mtDNA; circular DNA) and ribonucleic acid ('mitochondrial chromosome'). Several prokaryote-like features of the mitochondrion have led to revival of the proposition that it is a semi-autonomous structure of bacterial ancestry.

Mitosis   Process of nuclear division into two, accompanying cell division. Before the event, replication of the chromosomes (DNA replication) occurs so that the daughter cell nuclei possess chromosomal material identical with that of the parent cell.

See also MEIOSIS; NUCLEUS

Modification enzyme   Bacterial methylase(s) capable of methylating double-stranded DNA molecules at specific positions (specific nucleotide sequences), which makes the DNA resistant to hydrolysis by the specific endonuclease known as the restriction enzyme. The DNA of bacteriophages successful in infecting a given bacterial strain is found to have undergone modification in the same way as the host DNA.

See also RESTRICTION ENZYME

Molecular diseases   Various genetically determined abnormalities in humans associated specifically with deranged structure of macromolecules. Examples are the *hereditary anaemias*: 'sickle-cell anaemia' and β-thalassaemia ('Cooley's anaemia'), in which abnormal haemoglobins are produced because of defects in the gene (DNA) for globin synthesis. *Mucopolysaccharidoses* are clinical conditions in which dermatan sulphate and/or heparan sulphate accumulate deleteriously in lysosomes of cells owing to the genetic absence of a catabolising enzyme for these substances. In *Xeroderma pigmentosum* the defect is an inability to repair damage to DNA in skin fibroblasts.

Molecular hybridisation   *see* HYBRIDISATION TECHNIQUE

Monomer (Protomer)   A relatively simple molecular unit capable of joining with similar units to form a large molecule polymer. Amino acids are monomers of proteins and nucleosides are monomers of the nucleic acids. Proteins and nucleic acids can themselves be regarded as monomers in terms of their organisation within biological structures such as muscle or collagen fibres or chromosomes.

See also DIMER; NUCLEOSIDE AND NUCLEOTIDE

Muton   The smallest unit of genetic material capable of undergoing change (mutation)—deletion, replacement or additional insertion —i.e. one base pair in the polynucleotide chain.

Myoglobin (Muscle haemoglobin)   Protein akin to haemoglobin, with the same prosthetic group (haem), one to each molecule, having a reversible oxygen-combining storage function in vertebrate muscle. The molecule is a folded compact structure with about three-quarters of the chain comprising helical regions. The three-dimensional structure of sperm whale myoglobin has been analysed in detail. Approximate mol. wt. 17 000.

Myosin   Protein extracted from all types of muscle, forming about 60% of mammalian skeletal muscle. Mol. wt. about 500 000. Myosin molecules form the 'thick filaments' of skeletal muscle myofibrils, in which they are associated with the 'thin filaments' formed by molecules of actin. These two molecules partly overlap one another in a regularly repeated arrangement of contractile units (sarcomeres) along the myofibril. Subunits of globular myosin ('heavy meromyosin fraction') project sideways at (flexible?) junctions with the α-helical rod-like backbone ('light meromyosin fraction') of the filament along its length. These subunits form cross-links ('cross-bridges') with the actin molecules and are the force-generating structures responsible, possibly by a cyclic process of engagement and disengagement, for the contractile process, which according to the 'sliding filament model'[40, 48] involves movement of the actin and myosin filaments longitudinally past one another. This interaction of the molecules in vertebrate skeletal muscle involves adenosine triphosphatase activity (associated with the globular myosin) and is controlled by a functional unit formed by the protein complex troponin–tropomyosin[114] and calcium ions. The fibrous 'shaft' of the myosin molecule has been deduced[41] to be imperfectly helical in one region that consists of 200–300 amino acid residues; this shows special features and has been postulated as a 'hinge region' to explain the contractile properties of the molecule.

See also ACTIN

Nearest-neighbour sequence analysis Technique used to establish that the replication of deoxyribonucleic acid involves reproduction of the original nucleotide sequence in the copy strand[55]. Results are expressed as 'nearest-neighbour sequence frequency', i.e. the frequency with which a particular nucleotide is found next to a specific neighbour nucleotide in the polynucleotide chain.

See also REPLICATION

'Nick' Break in one strand (single-strand scission) of a deoxyribonucleic acid chain brought about by enzymic disruption of a phosphodiester bond, by action of an endonuclease.

Nodoc Name formed by reversing the word codon, suggested as a title for the codon's complementary base sequence that is present on a transfer RNA molecule.

Nonsense codon Nucleotide triplets (e.g. UAA, UAG) that do not code for amino acids. The codon UGA is a nonsense codon in *Escherichia coli* but might specify cysteine in vertebrate codes[11]. UAA and UAG are regarded as 'punctuation' codons, i.e. these codons interrupt the reading of the messenger RNA (mRNA) strand and also cause release of the synthesised polypeptide chain.

See also CODON; GENETIC CODE

28

N-Terminal   *see* AMINO TERMINAL

Nuclease   An enzyme capable of hydrolysing the polynucleotide molecule.

See also ENDONUCLEASES; EXONUCLEASES

Nucleic acids   Compounds of high molecular weights isolated from biological material (1871) and which on acid hydrolysis yield certain purine and pyrimidine bases, ribose sugar and phosphoric acid. Some nucleic acids (ribonucleic acids, RNA) yield D-ribose and others yield 2-deoxy-D-ribose (deoxyribonucleic acids, DNA); RNA and DNA also have different component bases. In all living organisms, including viruses, nucleic acids occur as aggregates (polynucleotides) of their monomer units (base–pentose–phosphate, called nucleotides or nucleoside phosphates). Individual nucleotides are linked via phosphodiester bonds to form the polynucleotide, i.e. each phosphate phosphorus links the hydroxyl groups of two ribose components, an arrangement that holds for both RNA and DNA. Dinucleotides (two monomers) can be prepared chemically and polynucleotides can be synthesised *in vitro* with enzymes (kinases and polynucleotide pyrophosphorylases).

See also BASE; NUCLEOSIDE AND NUCLEOTIDE

Nucleohistone   *see* CHROMATIN

Nucleolus   Dense spherical body or bodies visible within the resting nucleus of a cell and isolatable from nuclei by sonic disintegration. Consists of ribonucleoprotein, with a high concentration of ribonucleic acid (RNA), which accumulates after formation at the chromosomal DNA. Nucleoli are probably the site of formation of 45S RNA, the precursor of ribosomal RNA, and of methylation of transfer RNA[111].

Nucleoprotein   Protein held by electrostatic forces in association with deoxyribonucleic acid (DNA) of the nucleus. The low-molecular-weight protamines (nucleoprotamines) and histones (nucleohistones), both basic proteins, are isolated from cell nuclei and sperm-heads. Nucleoprotamines are thought to occupy the minor (shallow) groove of the DNA helix. Part at least of the nucleohistones occupy the major (deep) groove. Nucleohistones have been proposed to be repressors to the expression of certain tracts of the DNA in gene translation.

See also DOUBLE HELIX; GROOVE; HISTONES; PROTAMINES

Nucleoside and nucleotide   The monomeric unit of nucleic acids is the nucleotide (base–pentose–phosphate), generally named from the purine or pyrimidine base component with a prefix denoting the pentose sugar involved, i.e. ribo- or deoxyribo-. The nucleotide is a nucleoside monophosphate, i.e. the nucleoside is base–pentose. Major ribonucleosides (sugar component is ribose) and their recognised symbols are adenosine (A), cytidine (C), guanosine (G), ribosylthymine (T) and uridine (U); the analogous 2'-deoxyribonucleosides (sugar component is deoxyribose) are deoxyadenosine (2'-deoxyribosyladenine; dA), deoxycytidine (2'-deoxyribosylcytosine; dC), deoxyguanosine (2'-deoxyribosyl-guanine; dG) and deoxythymidine (thymidine; 2'- deoxyribo-sylthymine; dT).

Nucleus Discrete body of nucleoprotein (chromatin), usually spherical or oval, either enclosed by a lipoprotein membrane (eukaryotic cells) or without a membrane (prokaryotic cells), seen microscopically in the cells of most organisms. The nucleoprotein is in the form of long coiled threads which are capable of supercoiling and shortening, manifesting the typical thickened form of the cell chromosome(s) at the time of cell division. In the study of artificial hybrid cells (heterokaryotes) nuclei can be transferred from cell to cell *in vitro*.

See also MEIOSIS; MITOSIS; NUCLEOLUS

Ochre triplet Name given to the codon UAA after it was recognised that this terminal codon could be suppressed in the 'ochre' mutant of *Escherichia coli* bacteriophage T4.

Okazaki fragments (segments) Short-chain molecule of DNA first described in *Escherichia coli* by Okazaki *et al.* (1968)[86] and subsequently demonstrated in mammalian cell DNA. Synthesis of DNA occurs as discontinuous replication, the short chains first formed being combined enzymically by a polynucleotide ligase. Each DNA segment may be synthesised in *E. coli* from a primer of RNA[45, 118].

Oligomer A molecule resulting from the association of a few similar relatively simpler molecules. Most globular proteins are oligomers, formed by non-covalent assembly of 2–12 polypeptides (protomers or monomers).

Oligonucleotide Low-molecular-weight polynucleotide, comprising less than 20 nucleotide monomers, occurring in the mixture obtained on isolation of DNA or RNA.

Oncogene theory Hypothesis, now generally held to be unsubstantiated, that within all animal cell genomes is a set of genes (virogenes) capable of forming a C-type RNA virus, one of which (the oncogene) in certain circumstances can become activated so that the host cell becomes a cancer cell. Carcinogenic agents (including viruses) are held by the oncogene theory to have their effect by activating the oncogene[47].

Open-circle DNA *see* RELAXED CIRCULAR DNA

Operator A member gene of a set (operon) of 'structural' genes determining the amino acid sequence of a protein (enzyme), whose operation (transcription) is controlled by that gene[51]. The operator gene can be 'switched off' (repressed) by a repressor made by a 'regulator' gene.

Operon A bacterial or phage chromosome unit proposed[51] to explain the control or gene-dependent synthesis of proteins (enzymes). The operon comprises one or more 'structural genes' (each determining an amino acid sequence by its nucleotide sequence via mRNA) and an 'operator gene' controlling transcription of the structural gene(s) adjacent to the latter on the DNA strand. The '*lac* operon', regulating β-galactosidase production in *Escherichia coli*, and hence lactose-hydrolysing activity, was the first model studied.

See also INDUCTION; REPRESSOR

p Symbol for monosubstituted terminal phosphate residues used in abbreviated nucleotide nomenclature. For example, ppN and pppN are the diphosphate and triphosphate, respectively, of the nucleoside N. Recognised abbreviations for the nucleosides are: A, adenosine; C, cytosine; G, guanosine; T, ribosylthymine; U, uridine.

Paramyosin Fibrous protein of the adductor muscles of bivalve molluscs, thought[120] to form the core of a filament having molecules of myosin at the surface. By a phase change in the paramyosin the myosin–actin contraction system is held in the 'catch state', permitting the maintenance of muscle tension for a long time. Paramyosin molecules consist of two α-helix chains ('rope' formation).

See also MYOSIN

Peptide synthetase (Transferase I) Enzyme of the ribosome necessary for the formation of peptide bonds between amino acids in protein synthesis.

Permissive cell Host cell to a virus in which replication of the viral genome is allowed to take place after its integration into the chromosome(s).

Phage see BACTERIOPHAGES

Phase-shift mutation see FRAMESHIFT MUTATION

Phenotype Characteristic(s) exhibited by an organism, which might not be typical of the genes present (genotype), as for example as a consequence of the presence of dominant and recessive genes or an environmental effect.

Phosphodiester bond Link formed between the nucleotides of polynucleotide chains by covalent bonding of the phosphoric acid with the 3'-hydroxyl group of one ribose molecule and with the 5'-hydroxyl group of the next ribose ring.

See also POLARITY

Phosphodiesterase see EXONUCLEASES

Phytagglutinins (Lectins) Glycoproteins with the property of binding to the surface of transformed cells. For example, concanavalin A, a phytagglutinin used to study cell surface changes in cells transformed by DNA tumour viruses and by other agents.

Pitch The linear distance (3.4 nm) between two adjacent turns of the deoxyribonucleic acid double helix.

See also DOUBLE HELIX

Plasmid Genetic (deoxyribonucleic acid) element in bacterial cells that is additional to and independent of the cell chromosome and dispensable to the cell, capable of autonomous replication[65]. Examples are the F-factor ('sex factor'), R-factor ('resistance-transfer factor') and bacteriophages such as lambda phage.

See also FERTILITY FACTOR

β-Pleated sheet (β-Structure) Name given[89] to the major conformation of the fibrous protein molecule β-keratin. Molecular arrangement common in globular proteins, in enzymes such as lysozyme, ribonuclease, chymotrypsin, elastase, papain, carboxypeptidase, thermolysin and carbonic anhydrase, where the sheet is twisted (usually right-handed when viewed along the direction of the polypeptide chain)[16]. In some proteins the twisted pleated sheet is rolled into a cylinder (for example, chymotrypsin, elastase, thermolysin).

Plus strand Strand of 'parental' RNA in RNA bacteriophages that is used as a template for the production of a complementary RNA strand ('minus strand') with the formation of a double-stranded (double-helix) RNA ('replicative form').

Polarity   Orientation of a polynucleotide strand with respect to a partner strand expressed in terms of the internucleotide linkages. For example, in the DNA duplex the complementary chains are said to be of opposite polarity: in one strand the linkage is the 'normal' left to right arrangement, with the oxygen atom on C-3′ attached at the left of the diesterified phosphate residue and the oxygen atom on C-5′ attached at the right, i.e. C-3′→C-5′, and in the other it is right to left (C-5′→C-3′).

poly(A)   *see* POLYADENYLIC ACID

Polyadenylic acid [poly(A)]   Homopolymer sequences consisting of 50–200 adenine-containing nucleotides, i.e. poly(A), have been demonstrated in most messenger RNA (mRNA) molecules from eukaryotic cells[25, 26, 38] and animal viruses at the C-3′ end of the polynucleotides. This 'tail' of poly(A) is not translated and suggested functions for it include roles in cleavage maturation and in anchoring the protein associated with the mRNA as it is extruded from the nucleus to the cytoplasm and to polyribosomes[63]. In Hela cells the length of the poly(A) tract decreases with age[109] and has been reported in the cells' mitochondria[90]. In appropriate enzymic systems, *in vitro* poly(A) directs the synthesis of polylysine. Yeast mRNA contains a short poly(A) sequence[71].

Polycistronic messenger   Messenger ribonucleic acid carrying the amino acid sequence of several proteins in RNA viruses.

   See also CISTRON

Polymerase I and II   *see* RIBONUCLEIC ACID POLYMERASE

Polynucleotide kinase (Polynucleotide 5′-hydroxyl-kinase, EC 2.7.1.78)   An enzyme found in bacteriophage-infected *Escherichia coli* and in mammalian liver[49], capable of catalysing the transfer of phosphate from adenosine triphosphate to the 5′-hydroxy position in DNA nucleotides. A possible function in cells is to act with DNA ligase in the repair of single-strand breaks in DNA.

Polynucleotide ligase (Polynucleotide synthetase; EC 6.5.1.1 and 2)   Bacterial enzymes capable of catalysing the repair of breaks or 'nicks' in polynucleotide strands induced by endonuclease action or radiation injury. The ligases unite the free 5′-phosphate and free 3′-hydroxyl ends at the break. In the same way ligases are suggested to be responsible for growth in length of the polynucleotide chains by uniting short-chain fragments ('Okazaki fragments').

See also DEOXYRIBONUCLEIC ACID LIGASE; OKAZAKI FRAGMENTS

Polynucleotide phosphorylase (Polyribonucleotide nucleotidyltransferase, EC 2.7.7.8) Enzyme catalysing the synthesis of polyribonucleotide from nucleoside diphosphates[37]. A polymer with a random sequence of bases is produced from a mixture of nucleoside triphosphates in the absence of a template.

Polynucleotides Macromolecules formed by polymerisation, through phosphodiester bonds, of nucleotides; for example, the nucleic acids ribonucleic acid (RNA) and deoxyribonucleic acid (DNA). Polynucleotides can be prepared enzymically (deoxyribonucleic acid polymerase and polynucleotide phosphorylase) but chemical methods of synthesis yield only dinucleotides. Homopolymers of repeated single (monomeric) nucleotide sequences, e.g. polyadenylic acid [poly(A)] or polycytidilic acid [poly(C)], have been prepared, as have complexes of such homopolymers, e.g. poly(AC), and used to study the helical secondary structural form characteristic of these macromolecules[2].

See also NUCLEOSIDE AND NUCLEOTIDE; POLYADENYLIC ACID; POLYURIDYLIC ACID

Polypeptide Compounds of high molecular weight formed from large numbers of amino acids united by peptide bonds (—CO—NH—), i.e. whose main structure is generally representable as [—HN—CH(R)—CO—]$_n$.

See also PROTEIN

Polyploidy Cell state in which some multiple of the usual number (diploid) of chromosomes is present in the nucleus.

See also DIPLOIDY

Polyribosomes (Polysomes) Name given to the cluster of ribosomes temporarily associated with a single messenger RNA strand during protein synthesis in several different types of cells[57, 133].

See also RIBOSOMES

poly(U) see POLYURIDYLIC ACID

Polyuridylic acid [poly(U)] A homopolymer of uracil-containing nucleotides. Synthesised poly(U)[2] in an appropriate enzymic system *in vitro* directs the synthesis of polyphenylalanine[83, 84].

Primary structure Basic components of a macromolecule and their sequential arrangement. Bonding in primary structures is predominantly covalent. *Protein primary structure*: the amino acid composition and linear sequence and the number of polypeptide

34

chains in the molecule[68]. *Polynucleotide primary structure:* the number and sequence of nucleotides in the molecule.

See also SECONDARY STRUCTURE; QUATERNARY STRUCTURE; TERTIARY STRUCTURE

Primer Name given to the polynucleotide chain required in enzymic systems *in vitro* to act as template for the synthesis of the complementary chain.

See also DEOXYRIBONUCLEIC ACID POLYMERASE

Procollagen Higher molecular weight form of collagen found intracellularly and believed to be the precursor of this protein.

See also COLLAGEN

Prokaryote (Prokaryotic cell; Prokaryocyte) An organism with cells appearing to be without a discrete nuclear membrane enclosing the genetic material. Bacteria and blue–green algae are prokaryote organisms[117].

See also EUKARYOTE

Promoter site Position on the deoxyribonucleic acid (DNA) helix at which ribonucleic acid (RNA) polymerase interacts to initiate transcription (RNA synthesis)[29, 52]. The site requires to be activated before this occurs (in *Escherichia coli*) by contact with cyclic AMP and CRP (a protein).

See also CYCLIC AMP

Prophage Bacteriophage genetic material carried as part of the infected host cell's genome, either at a specific site or in other types at several sites.

See also BACTERIOPHAGES; LYSOGENY

Protamines Basic proteins occurring in association with DNA (nucleoprotamine) in cell nuclei. Salmon sperm salmine is a much-studied protamine. The function of these proteins is unknown.

Protein Large molecules (mol. wt. usually above 10000), universally present in living organisms and yielding α-amino acids on hydrolysis enzymically or with acid or alkali. Those yielding only amino acids are *simple proteins*, those yielding compounds in addition to amino acids are *conjugated proteins*. Controlled hydrolysis can yield polypeptides and peptides. The amino acids are united by peptide bonds (—CO—NH—) and the peptide chains can also be bridged by disulphide links of cystine residues. Amino acid sequences of proteins have been determined by end-group analysis and the three-dimensional conformation has been derived for many proteins by x-ray diffraction analysis. Fibrous and globular proteins can be distinguished; keratin and collagen are

35

examples of fibrous proteins and plasma proteins, haemoglobin and enzymes are typical globular proteins. The natural conformation of proteins, involving set patterns of folding or coiling of the polypeptide chains, is changed in the process known as denaturation, when the structure becomes disordered and the physical and chemical properties are altered. The range of molecular weights of proteins is shown in *Table 2*.

See also PROTEIN SYNTHESIS

Table 2 RANGE OF MOLECULAR WEIGHTS OF PROTEINS

Insulin	Cow	6 300
Cytochrome c	Cow	12 400
Virus coat protein	Tobacco mosaic virus	18 270
Trypsin	Cow	23 800
Carbonic anhydrase		31 000
Egg albumin (ovalbumin)		44 000
Serum albumin	Human	65 000
Haemoglobin	Human	66 000
Actin		70 000
Hexokinase		90 000
γ-Globulin	Cow	180 000
	Human	185 000
Catalase		250 000
Serum fibrinogen	Cow	330 000
	Human	450 000
Myosin		500 000
β-Galactosidase	*Escherichia coli*	540 000
Thyroglobulin		630 000
Ferritin		747 000
Actomyosin	Rabbit	5×10^6
Haemocyanin	Snail *(Helix pomatia)*	9×10^6

Protein synthesis Protein-synthesising information in the form of messenger RNA (mRNA) arises from the cell nucleus and is acted upon at the ribosomes. Stages recognised are: (1) initiation, (2) elongation and (3) termination[69]. Synthesis is initiated by the formation of an 'initiation complex' of a ribosome subunit, mRNA and initiator transfer RNA (tRNA). In eukaryotes the initiator tRNA is methionyl tRNA $_f^{112}$ (Met-tRNA$_f$) and in prokaryotes it is N-formylmethionyl tRNA $_f^{66}$ (fMet-tRNA$_f$). The first peptide bond is formed, the amino terminal residue being the first unit of the polypeptide chain that is to be built up from amino acids joined in the order specified by the mRNA. This and subsequent additions of amino acids to form the chain take place on the ribosome. In *Escherichia coli*, codons AUG and GUG place formylmethionine at the start of the polypeptide, and the 'nonsense codons' UAA, UAG and UGA form the codon signals for chain termination, when the carboxyl terminal of the polypeptide is completed.

See also RIBOSOMES; TRANSFER RNA

Protovirus theory Hypothesis that a messenger ribonucleic acid ('protovirus RNA') can travel from cell to cell in the animal body, providing a means of intercellular communication with the possibility of leaving a special deoxyribonucleic acid sequence within each invaded cell[123]. Embryogenesis, immunological instructions and memory have been suggested to involve the protovirus.

Provirus A double-stranded DNA copy of the RNA tumour virus genome, representing the genetic material of the virus within an infected (transformed) host cell and thought to be the template for production of new virus genomes[123, 124].

See also TRANSFORMATION

Psi factor *see* CYCLIC AMP RECEPTOR PROTEIN

Puff An area of enlargement in diameter, observed in 'giant' chromosomes of insects, that occurs in a specific position (chromomere) in which gene activity (transcription of deoxyribonucleic acid) is taking place.

See also CHROMOSOME

Purine bases Derivatives in a chemical sense of the heterocyclic double ring compound purine. Purine bases characteristically found as components of nucleic acids are adenine (6-aminopurine) and guanine, with their N-methyl- and C-methyl-substituted derivatives occurring as the 'rare purine bases' in small amounts. Adenine and guanine occur in both deoxyribonucleic acid and ribonucleic acid.

See also ADENOSINE; GUANOSINE

Pyrimidine bases The heterocyclic compound pyrimidine is the parent substance for the bases uracil (2,6-dihydroxypyrimidine), cytosine (2-hydroxy-6-aminopyrimidine), thymine (2,6-dihydroxy-5-methylpyrimidine) and, less characteristically, 5-methylcytosine, 5-hydroxymethylcytosine, 1-methyluracil and 1-methylcytosine, occurring in nucleic acids. Uracil, cytosine and their 1-methyl derivatives occur in ribonucleic acid; cytosine, thymine and 5-methylcytosine and 5-hydroxymethylcytosine occur in deoxyribonucleic acid.

See also CYTIDINE; RIBOSYLTHYMINE; URIDINE

Quaternary structure Arrangement and interrelationships of different kinds of biological macromolecules, particularly proteins, in subcellular aggregates or tissue units such as cell membranes, ribosomes, muscle or nerve fibres. Most recently these structures have been analysed by a neutron-scattering technique.[27]

See also PRIMARY STRUCTURE; SECONDARY STRUCTURE; TERTIARY STRUCTURE

Random coil (Cyclic coil) One of the irregularly coiled forms in which polymeric macromolecules can occur in solution (in contrast with 'rigid coil', spherical or rod-like forms). Denatured fibrous protein molecules form random coils. Ribonucleic acid in solution appears to be a random coil (by light-scattering properties[139]) and so is the denatured DNA formation, when the two strands of the double helix are separated.

See also ANNEALING; DENATURATION

'Reading' One-way linear process by which amino acid sequences are recognised from the nucleotide code constituted by messenger RNA (mRNA) by cellular protein-synthesising systems. The direction of reading is the same as the direction of synthesis of the mRNA chain in transcription, from the C-5' end to the C-3' end.

See also PROTEIN SYNTHESIS; TRANSLATION

Recombination An apparently universal process producing transfer of genetic material, either as an exchange between chromosomal segments in the same nucleus during cell division or as a movement from donor cell to recipient cell (in bacteria). Spontaneous breakage and recombination of chromosomes by crossing over, so that each recombinant chromosome incorporates segments from both of its homologous parental chromosomes, has been used to develop gene maps for a number of organisms.

Recon The smallest unit of genetic material within a gene (cistron) that is capable of recombination, i.e. one base pair in the polynucleotide chain.

Regulator gene *see* REPRESSOR

Regulatory site *see* ALLOSTERIC EFFECTOR

Regulon System of genes for the induction of enzymes in bacterial cells, whose activity is controlled by a single repressor substance[72].

38

A regulon can be formed by either a single operon or several operons.

Reiteration (Repetitive sequences) Existence of 'copies' of a DNA sequence, i.e. repetitive nucleotide sequences, in a cell genome[12, 136]. Reiteration frequencies of various specific DNA sequences have been estimated; for example, two or three reiterated genes for duck haemoglobin[8].

Relaxed circular DNA (Open-circle DNA) Covalently joined polynucleotide double-stranded ring structure occurring in bacterial cells in which one strand is 'nicked', so that the circle of that strand is broken. This is the form of the bacteriophage φX 174 RF 2 (replicative form 2) DNA. Replication by the 'rolling circle model' can occur with the closed strand as template.

Repair mechanisms Cellular enzyme systems capable of restoring DNA damaged by radiation or antibiotics. Such damage may involve distortion of the DNA helix by dimer formation (u.v.-radiation) or the creation of breaks (x-irradiation, etc.). 'Excision–patch' ('cut and patch') models of the repair process have been proposed, in which after recognition of a defect in the DNA, 'nicks' are made in the chain by endonuclease action ('repair enzyme') to cut out the abnormal fragment; polymerisation then occurs by DNA polymerase activity to replace the excised fragment and finally polynucleotide ligase joins up the new section to the old.

Repetitive sequences *see* REITERATION

Replacement The exchange of one base pair for another in the double-stranded DNA helix, responsible for errors in transcription. The exchange can involve either transition, in which the exchanged purine–pyrimidine pair are each allocated to the same strand as the original, or transversion, in which a purine takes up the pyrimidine position and its complementary pyrimidine is inserted in the other strand[30]. Replacement of bases is a method of inducing mutations.

See also DELETION; INSERTION; TRANSCRIPTION

Replicase *see* RIBONUCLEIC ACID–DEPENDENT RNA POLYMERASE

Replication Process of duplication of each of the two strands of deoxyribonucleic acid (DNA), a new complementary strand being formed on each parent strand. In viruses the single-stranded ribonucleic acid (RNA) first gives rise to complementary ('minus') strands and these are replicated to form the 'plus' strand form of the virus RNA.

Replicon hypothesis Theory that initiation of chromosome replication involves a 'replicator', located on the DNA, which is acted upon by a factor called an initiator.

See also INITIATION FACTORS

Repression   The stopping or inhibition of synthesis of an enzyme by a cell, taking place when the essential product of the enzyme's activity accumulates or is provided. A regulatory gene for the enzyme controls the structural gene for the enzyme by directing production of a repressor molecule, which binds to the gene operator and causes cessation of the structural gene's action[51].

Repressor   Protein synthesised by a regulator gene, which by binding to a specific site on DNA (the operator gene of an operon) prevents the formation of messenger RNA by the operon's other (structural) genes and hence stops protein (enzyme) synthesis[51]. Inactivation of repressor involves its combination with an effector or inducer.

See also COREPRESSOR; INDUCTION; OPERON

Restriction enzyme   Bacterial endonuclease(s) capable of recognising and hydrolysing double-stranded DNA foreign to the cell. Invading DNA is 'nicked' by breakage of phosphodiester bonds at specific regions of the molecule (specific nucleotide sequences), which are protected in the host DNA by methylation. Deoxyribonucleases further hydrolyse DNA nicked by restriction enzyme(s).

See also MODIFICATION ENZYME

Reverse transcriptase   *see*   RIBONUCLEIC ACID-DEPENDENT DNA POLYMERASE

Rho factor   A protein responsible for the termination of transcription at specific sites on bacteriophage lambda DNA[99] and bacteriophage fd-RF DNA[122]. It is one of the termination factors.

Ribonuclease (RNase; RNAase)   Enzyme that degrades ribonucleic acid (RNA), found in most cells. Pancreatic ribonuclease[62] (ribonuclease I; ribonucleate 3'-pyrimidino–oligonucleotidohydrolase: EC 3.1.4.22) and *Escherichia coli* ribonuclease (ribonuclease II; ribonucleate 3'-oligonucleotidohydrolase: EC 3.1.4.23) have been particularly studied. Site of ribonuclease action on the RNA molecule is the phosphodiester bond, the ribonucleotide being degraded to mononucleotides or small groups of nucleotides; it is thus a phosphodiesterase specific for RNA, and an endonuclease.

Ribonucleic acid (RNA)   A polymer of ribonucleotides isolated from cells in association with protein as ribonucleoprotein. RNA nucleotides have the pyrimidine base uracil instead of the base thymine which is present in deoxyribonucleic acid (DNA). RNA

40

occurs in cells in three major forms, all predominantly extranuclear, differing in molecular-weight ranges and component nucleotides: over 80% is rRNA (ribosomal RNA, q.v.), about 15% is tRNA (transfer RNA, q.v.) and about 5% is mRNA (messenger RNA, q.v.). Nuclear RNA (nRNA) of all three types is recognised and RNA is probably synthesised in the nucleus by transcription of one strand of DNA. A large proportion of nuclear RNA occurs in the nucleoli. Isolated RNA fractions are commonly denoted by their sedimentation coefficients, e.g. '23S RNA' and '16S RNA' from *Escherichia coli*. In different RNA species the molar proportions of bases present differ. The nucleotides are linked by C-3' to C-5' phosphodiester bonds to form single-strand polynucleotide chains in which are present short imperfectly helical regions with base pairing by hydrogen bonds between adenosine and uridine and between guanosine and cytidine. The role of RNA in cellular protein synthesis is well established. RNA is said to be involved in the initiation of DNA synthesis in *E. coli*[118]. RNA has been postulated to be closely concerned with memory storage and learning in animal nervous systems. 'Infective RNA' is the cell-infecting agent of some viruses and bacteriophages. Viral RNA shows special features in its primary and secondary structure.

 See also MESSENGER RNA; RIBOSOMAL RNA; TRANSCRIPTION; TRANSFER RNA

Ribonucleic acid (RNA) bacteriophages Simplest known viruses, with bacteria as hosts. Genetic material of a bacteriophage particle is a single-stranded RNA molecule; this is surrounded by 180 identical molecules of coat protein with a second protein (perhaps only one molecule of this), called the maturation factor (A-protein), associated with attachment of the bacteriophage to the host cell. In the host, the bacteriophage RNA is treated as messenger RNA (mRNA), translated by the host with production of phage proteins. Production of RNA polymerase as one of these promotes replication of an RNA molecule (minus strand) complementary to the original phage RNA (plus strand). RNA polymerase synthesises 'progeny plus strands' from the 'minus strand' template. These strands in turn act as mRNA to be translated into further phage proteins, finally with association to give new progeny bacteriophage particles.

Ribonucleic acid (RNA)-dependent DNA polymerase (Reverse transcriptase; Temin's enzyme) An enzyme first isolated from RNA tumour virus particles[31]. In cells infected by the virus the

41

enzyme synthesises provirus DNA by use of the virus single-stranded RNA genome as a template. Since the purified virus enzyme can also employ other natural and synthetic RNA molecules in this way it has been used extensively for the synthesis of complementary DNA in experimental systems. A similar enzyme in human leukaemic cells has been reported[125].

See also PROVIRUS; TRANSFORMATION

Ribonucleic acid-dependent RNA polymerase (Replicase; Synthetase) Enzyme occurring in micro-organisms and vertebrate cells, capable of initiating replication of RNA. In RNA virus-infected cells a high activity of the enzyme is present.

Ribonucleic acid (RNA) polymerase (Nucleoside triphosphate–RNA nucleotidyltransferase, EC 2.7.7.6; DNA-dependent RNA polymerase; Polymerase I and II; RNA nucleotidyltransferase; Transcriptase) Enzyme of mammalian cell nuclei responsible for synthesising RNA by transcription, DNA being used as template. Two major forms of the enzyme are recognised [14, 101]: form A (or I) synthesises ribosomal RNA (rRNA), preferentially using native DNA as template and occurring in the nucleolus; form B (or II) synthesises messenger RNA (mRNA) or heterogeneous nuclear RNA, denatured DNA being the preferred template, and occurs in the nucleoplasm. The various forms have been postulated to regulate transcription by binding selectively to different regions ('promoter sites') of DNA.

See also PROMOTER SITE

Ribonucleic acid replicase *see* RIBONUCLEIC ACID–DEPENDENT RNA POLYMERASE

Ribonucleic acid synthetase *see* RIBONUCLEIC ACID–DEPENDENT RNA POLYMERASE

Ribonucleic acid (RNA) viruses *see* VIRUS

Ribonucleoside and ribonucleotide *see* NUCLEOSIDE AND NUCLEOTIDE

Ribose Pentose sugar present in all cells as a component of nucleic acid in polyribonucleotides. Ribose phosphate molecules linked by phosphodiester bonds form chains that make the backbone of polyribonucleotides, as in RNA.

Ribosomal DNA (rDNA) DNA responsible for the specification of ribosomal RNA (rRNA) structure, i.e. DNA complementary to rRNA.

Ribosomal RNA (rRNA) Major form of RNA (over 80%) in cells, found in ribosomes and of high molecular weight (in different species from 0.6×10^6 to 2.1×10^6). rRNA is synthesised in the cell nucleus by transcription from defined regions of DNA and in mammalian cells undergoes 'maturation' before its appearance in the ribosomes. Evidence suggests that rRNA is not translatable, i.e. it does not carry protein sequence information.

See also RIBONUCLEIC ACID; RIBOSOMES

Ribosomes   Subcellular particles, approximately 20 nm in diameter, found in all types of organisms. They consist of proteins (40%) and RNA (60%). Dissociation of ribosomes produces two. subunits, usually referred to by their sedimentation coefficients: 30S and 50S subunits from *Escherichia coli* ribosomes and 40S and 60S subunits from mammalian ribosomes. In association with 'initiation factors' ribosomes are concerned with protein synthesis. The smaller subunit engages with the start position of the messenger RNA (mRNA)[66, 69], forming the 'initiation complex' together with initiator transfer RNA (tRNA). tRNA molecules can bind at two sites on the larger subunit of the ribosome: site A is for aminoacyl-tRNA and site P for peptidyl-tRNA. During a temporary linkage of the tRNA anticodon with the appropriate codon of the mRNA, the amino acid attached to the tRNA at site A is peptide-bonded to the forming polypeptide chain held by tRNA at site P, the enzyme peptidyl transferase of the ribosome being responsible. As the ribosome moves to the next codon of the mRNA one free tRNA is released and a new aminoacyl-tRNA joins at site A, to supply the next amino acid required according to the template. Other ribosomes can engage with the mRNA as the start position is available, and the strand of several ribosomes is known as the polysome.

See also POLYRIBOSOMES; PROTEIN SYNTHESIS; TRANSFER RNA

Ribosylthymine (Ribothymidine)   One of the deoxyribonucleosides (containing the pyrmidine base thymine) making up deoxy-ribonucleic acid (DNA) and forming one unit of the genetic code (T). Ribosylthymine, which is not present in ribonucleic acid (RNA), 'pairs' non-covalently with adenosine (A) in DNA.

See also BASE

Ribothymidine   *see* RIBOSYLTHYMINE

RNA and terms prefixed by RNA   *see* RIBONUCLEIC ACID

Rolling circle model   Postulated method of replication of bacteriophage circular DNA[33] in which the closed inner (minus strand) ring serves as 'endless' template for the elongation of the 'nicked' outer (plus strand) at the 3′-hydroxyl end by addition of nucleotides, whilst the 5′-phosphate end is anchored.

rRNA   *see* RIBOSOMAL RNA

Satellite DNA   A minor fraction of nuclear DNA from eukaryotic organisms, separated by isopycnic caesium chloride centrifugation[60], composed of numerous tandemly repeated nucleotide sequences. Function is unknown, although this fraction is thought to represent non-coding stretches of the genome.

Scripton   *see* TRANSCRIPTON

Secondary structure   Molecular conformation of biological macromolecules[68]. The helical arrangement, with hydrogen bonding, present in both proteins and polynucleotides is the secondary structure of these molecules.

   See also PRIMARY STRUCTURE; QUATERNARY STRUCTURE; TERTIARY STRUCTURE

Semiconservative replication   Replication (duplication) of double-stranded DNA involving separation of the parental strands and the formation of a newly synthesised complementary strand for each. The parental strands are conserved although their association is not[79].

Sense strand (Plus strand)   The single DNA strand of the double helix which is complementary to the RNA formed by transcription. Its partner strand, of which the formed RNA must be a copy, is the 'antisense strand'.

   See also TRANSCRIPTION

Sex factor (bacterial)   *see* FERTILITY FACTOR

S factors (S_1, S_2, S_3)   Proteins isolated from bacteria and shown to have a role   in lengthening of the polypeptide chain in protein synthesis ('elongation factors').

Sigma factor   Polypeptide subunit of bacterial RNA polymerase important for initiation of transcription by selection of activated promoter sites[35] on the DNA template. This was the first of a number of agents to be postulated as transcription factors and 'sigma factors' is used as a generic term for these.

   See also INITIATION FACTORS

Single-stranded DNA   This form of DNA occurs in the bacterial virus φX 174 (host *Escherichia coli*)[110]. This DNA species has 5500 nucleotides and the strand is covalently closed to form a ring ('circular or cyclic DNA').

Soluble RNA (sRNA)   Obsolete term. *See* TRANSFER RNA

'Sticky ends'   *see* COHESIVE TERMINUS

Supercoil   *see* SUPERHELIX

Superhelix (Coiled coil; Supercoil structure)   (1) Macromolecular structure formed from a number of α-helical polypeptide strands

44

twisted together, as seen in fibrous proteins[88]. (2) Helical double helix: fibre of 20–30 nm diameter formed (chromatin) by twisting of the DNA double helix (2 nm diameter)[18, 24]. (3) Tertiary molecular structure formed from the duplex helical DNA fibre in its circular (cyclic) form by a number of twists in the covalently closed ring. This is the form of the bacteriophage φX174 RF 1 (replicative form 1) DNA, which is converted into the RF2 open or relaxed circle form by a single-strand scission ('nick').

See also WINDING NUMBER

Suppression (Frameshift suppression)　Restoration of the correct translating 'reading frame' after occurrence of a frameshift mutation, either by a compensatory additional frameshift (internal suppression[21]) in the nucleotide chain or by involvement of transfer RNA (suppressor tRNA: external suppression[97]) to maintain the proper phasing.

See also FRAMESHIFT MUTATION; TRANSLATION

Synonymous codon　*see* DEGENERATE CODE

Synthetase　*see* RIBONUCLEIC ACID-DEPENDENT RNA POLYMERASE

Temin's enzyme　*see* RIBONUCLEIC ACID-DEPENDENT DNA POLYMERASE

Template　Pattern for polynucleotide synthesis, represented by a specific polynucleotide molecule, for example deoxyribonucleic acid. In protein synthesis the nucleotide sequences coding for amino acids can be regarded as a template for the forming protein.

Termination factors　Cellular components required for recognition of the (termination) site on a template molecule for cessation of the process of transcription or protein synthesis. For example, termination factors R1 and R2 of *Escherichia coli* recognise the chain-terminating codons UAA, UAG and UGA.

See also KAPPA FACTOR; RHO FACTOR

Terminator codons　Triplet nucleotides (codons) acting as 'full stops' in the code sequence for synthesis of a protein, i.e. as signals for stopping the elongation of the polypeptide chain.

Tertiary structure　Three-dimensional arrangement of the components of biological macromolecules[68]. Hydrophobic and van der Waals forces are characteristically involved in these arrangements. Loss of biological activity occurs with loss of tertiary structure; for example, reduction of pancreatic ribonuclease leads to unfolding of the molecule of the enzyme and loss of activity, reversible by oxidation.

See also PRIMARY STRUCTURE; QUATERNARY STRUCTURE; SECONDARY STRUCTURE

T factors (T$_s$, T$_u$)   Proteins isolated from bacteriophage and bacteria and shown to have a role in lengthening of the polypeptide chain in protein synthesis ('elongation factors').

Thymidine   *see* NUCLEOSIDE AND NUCLEOTIDE

Thymine (2,6-Dihydroxy-5-methylpyrimidine)   A derivative of the heterocyclic compound pyrimidine, forming a principal base in nucleic acids.

See also BASE; RIBOSYLTHYMINE

Toroid structure   One suggested form for the secondary molecular structure of the superhelical circular duplex DNA found in the replicative form of bacteriophage ΦX174. *Toroid* (Webster's Dictionary): a surface generated by a plane closed curve rotated about a line in its plane that does not intersect the curve.

Transcriptase   *see* RIBONUCLEIC ACID (RNA) POLYMERASE

Transcription   Process by which deoxyribonucleic acid (DNA) gives rise to ribonucleic acid (RNA), one strand only of the duplex DNA serving as template for the new polynucleotide. The bases of the RNA are complementary to those of the DNA strand, just as is found in the DNA duplex, with the exception that uracil takes the place of thymine in base pairing with adenine. Thus, bases adenine, guanine, thymine and cytosine in DNA give rise to bases uracil, cytosine, adenine and guanine, respectively, in RNA. RNA formed in this way to act as a programme for synthesis of protein is known as messenger RNA (mRNA).

See also COMPLEMENTARY TRANSCRIPTION

Transcripton (Scripton)   Operational unit postulated[121] for the transcription process in which a single autonomous promoter defines the unit, capable of synthesising several messenger RNA molecules at various rates. The operon is a simple transcription system.

Transduction   Process of transfer of genetic material (DNA or RNA) from bacteriophages to bacteria or passage of such material between bacteria with 'transducing bacteriophage' as the intermediary.

Transfer RNA (tRNA) (Acceptor RNA; Adaptor; Dictionary RNA; Soluble RNA; sRNA)   A low-molecular-weight (23 000–28 000) RNA concerned with amino acid transport in cells and possibly with maintenance of the correct 'reading frame' translation[97]. The molecule is single stranded with loops and folds involving short imperfect helical regions of base-pairing. At least one form of tRNA exists specifically for each amino acid, for example alanine–tRNA (abbreviated as tRNA^{Ala}). Approximately constant in length in different species (about 75 nucleotides), the tRNA chain frequently shows the nucleoside guanosine at the C-5′ terminal end and always shows cytidine–cytidine–adenine (CCA) at the C-3′ terminal (acceptor) end where the amino acid becomes attached (to A). In its mid-length the tRNA chain has three base groups forming the anticodon, which links with the corresponding triplet (codon) in messenger RNA (mRNA) during protein synthesis at the ribosome. A two-dimensional structure for tRNA that is consistent with its known nucleotide sequences and giving maximum base-pairing is known as the 'cloverleaf pattern'[46]. tRNA 'recognises' the specific aminoacyl–tRNA synthetase and accepts the appropriate activated amino acid, for example alanyl–tRNA (Ala–tRNA). The temporary anticodon–codon linking with the mRNA ensures that the amino acid carried by the tRNA is placed in correct sequence in the protein being synthesised. Electron-density studies have revealed that, three-dimensionally, this pattern has an L shape with the amino acid receptor at the end of one limb and the anticodon at the end of the other[59].

See also ANTICODON; RIBONUCLEIC ACID; WOBBLE HYPOTHESIS

Transferase I   see PEPTIDE SYNTHETASE

Transferase II   see TRANSLOCASE

Transformation   The change in a cell by which a new genetic trait becomes established in the cell line after infection by a tumour virus. Postulated to be the result of integration of DNA complementary to the virus RNA into the host cell DNA[123]. Cells transformed by tumour-producing viruses may show abnormal growth: they change their shape, growth rate is accelerated and in tissue culture the cells do not adhere to surfaces like normal cells.

See also   PROVIRUS;   RIBONUCLEIC   ACID-DEPENDENT   DNA POLYMERASE

Transition   see REPLACEMENT

Translation   Process occurring in the early stage of synthesis of protein, by which the nucleic acid code of messenger RNA (mRNA) is 'read' as codons for amino acids in the order of their incorporation into the polypeptide chain. Ribosomes are the cytoplasmic sites for translation and for the synthesis of protein from 'activated' amino acids delivered to them attached to specific transfer RNA (tRNA) molecules, anticodons to the mRNA.

See also ACTIVATION; ANTICODON; CODON

47

Translocase (G factor; Transferase II) Enzyme (complex), one of the 'elongation factors', postulated to catalyse the process of translocation on ribosomes during protein synthesis in *Escherichia coli*[85]. Guanosine triphosphate (GTP) is associated with the enzyme's activity.

Translocation Movement of molecule or molecular complex from one functional site in a subcellular particle to another, postulated to be initiated by enzyme activity ('translocase'). For example, in ribosomal protein synthesis in *Escherichia coli*, formation of the peptide bond is accompanied by translocation of the transfer RNA from site A to site P.

See also RIBOSOMES

Transversion *see* REPLACEMENT

TR factor (Transfer RNA releasing factor) Enzyme postulated[50] to be involved in release of transfer RNA from the ribosome during protein synthesis in *Escherichia coli*.

Trimer *see* CODON

Trinucleotide *see* CODON

Triplet hypothesis *see* CODON

Triploidy The condition in which three sets of chromosomes, instead of the normal two sets (diploidy), per cell are present. Viable organisms showing this state have been reported in amphibians, insects and plants, although mammalian triploid animals, including man, seldom survive birth.

tRNA *see* TRANSFER RNA

'Unwindase' (Albert's protein; T4 gene 32 protein) An enzyme in bacteriophage T4 postulated to be responsible for the denaturation of double-stranded DNA. By binding the DNA it causes the strands to unwind and aligns the two separated parental template strands so that the molecule is replicated by the activity of DNA polymerase(s).

Uracil (2,6-Dihydroxypyrimidine) A derivative of the heterocyclic compound pyrimidine, forming a principal base in ribonucleic acid (RNA).

See also BASE; URIDINE

Uridine One of the ribonucleosides (containing the pyrimidine base uracil) making up ribonucleic acid (RNA) and forming one unit of the genetic code (U). Uridine 'pairs' non-covalently with adenosine in RNA.
See also BASE

Virion The transmissible virus particle as visualised by electron-microscope techniques. A spherical virion consists typically of an outer glycoprotein–lipid layer surrounding an inner protein membrane that encloses a core of nucleoprotein (RNA– or DNA–protein), the core being about two-thirds of the diameter (80–120 nm) of the whole.
See also VIRUS

Virus Living matter of ultramicroscopic size capable of invading living cells and self-replicating (reproducing) only within such host cells. All types of plants and animals form hosts to specific viruses; the bacterial viruses known as bacteriophages ('phages') have been widely studied by molecular geneticists. Viruses can be classified from the nucleic acid of their genomes as DNA viruses (e.g. herpes simplex, polyoma and vaccinia viruses) or RNA viruses (e.g. f2 bacteriophage, influenza, mumps and tobacco mosaic viruses). In the latter class, RNA is replicated from RNA without DNA being involved. A proposed third class is the RNA–DNA viruses in which the genome is alternately RNA and DNA. The Rous sarcoma viruses (leucoviruses) are examples of this third type, with the unusual feature of DNA being translated from RNA.

Watson–Crick model *see* DOUBLE HELIX
Winding number (α; Linkage number) Number of times one strand is turned about the other in a closed circular duplex DNA molecule that is constrained to lie in a plane[131]. For example, for the polyoma virus DNA α = +445 (+ indicates clockwise or right-hand winding).
Wobble hypothesis Suggested by Crick[19] to explain the fact that transfer ribonucleic acid (tRNA) anticodons, when pairing with

messenger ribonucleic acid (mRNA) codons, show rigorous specificity for two of the three bases in each codon but will accept a variety of bases in the third position. For example, the phenylalanine tRNA anticodon AAG will recognise either UUU or UUC codons; generally, U recognises either A or G, G recognises U or C and I recognises U, C or A. The hypothesis states that variation ('wobble') in the codon's third nucleotide base is acceptable within defined limits imposed by rules of base pairing.

Zimm plot A graphical method[139] of interpreting analytical results from light-scattering measurements of biological macromolecules to determine molecular weight and dimensions.

References

1. Adams, J. M., Jeppesen, P. G. N., Sanger, F. and Barrell, B. G. *Nature, Lond.*, **223**, 1009–1014 (1969)
2. Arnott, S., Fuller, W., Hodgson, A. and Prutton, I. *Nature, Lond.*, **220**, 561–564 (1968)
3. Astbury, W. T. and Street, A. *Phil. Trans. R. Soc. Ser. A*, **230**, 75 (1931)
4. Beard, P. *Biochim. biophys. Acta*, **269**, 385–396 (1972)
5. Benbow, R. M., Eisenberg, M. and Sinsheimer, R. L. *Nature, Lond., New Biol.*, **237**, 141–143 (1972)
6. Benzer, S. *Prov. natn Acad. Sci. U.S.A.*, **41**, 344–354 (1955)
7. Bird, R. E., Lovarn, J., Martuscelli, J. and Caro, L. *J. molec. Biol.*, **70**, 549–566 (1972)
8. Bishop, J. O. and Rosbash, M. *Nature, Lond., New Biol.*, **241**, 204–207 (1973)
9. Blattner, F. R., Dahlberg, J. E., Boettiger, J. K., Fiandt, M. and Szybalski, W. *Nature, Lond., New Biol.*, **237**, 232–236 (1972)
10. Braunitzer, G., Best, J. S., Flamm, U. and Schrank, B. *Hoppe-Seyler's Z. physiol. Chem.*, **347**, 207–211 (1968)
11. Brenner, S., Barnett, L., Katz, E. R. and Crick, F. H. C. *Nature, Lond.*, **213**, 449–450 (1967)
12. Britten, R. J. and Kohne, D. E. *Science*, **161**, 529–540 (1968)
13. Brutlag, D. and Kornberg, A. *J. biol. Chem.*, **247**, 241–248 (1972)
14. Chambon, P., Gissinger, F., Mandel, J. L., Jr, Kedinger, C., Gniaydowski, M. and Meilhac, M. *Cold Spring Harb. Symp. quant. Biol.*, **65**, 693–707 (1970)
15. Chargaff, E. In *The Nucleic Acids*, Vol. 1 (ed. by Chargaff, E. and Davidson, J. N.), Academic Press, New York (1955)
16. Chothia, C. *J. molec. Biol.*, **75**, 295–302 (1973)
17. Clark, B. F. C. and Marcker, K. A. *J. molec. Biol.*, **17**, 394–406 (1966)
18. Crick, F. H. C. *Nature, Lond.*, **170**, 882–883 (1952)
19. Crick, F. H. C. *J. molec. Biol.*, **19**, 548–555 (1966)
20. Crick, F. *Nature, Lond.*, **234**, 25–27 (1971)
21. Crick, F. H. C., Barnett, L., Brenner, S. and Watts-Tobin, R. J. *Nature, Lond.*, **192**, 1227–1232 (1961)
22. Delange, R. J., Fambrough, D. M., Smith, E. L. and Bonner, J. *J. biol. Chem.*, **243**, 5906–5913 (1968)
23. Delange, R. J., Fambrough, D. M., Smith, E. L. and Bonner, J. *J. biol. Chem.*, **244**, 319–334, 5669–5679 (1969)

24. DuPraw, E. J. *DNA and Chromosomes*, Holt, Rinehart and Winston, New York (1970)
25. Edmonds, M. and Abrams, R. J. *biol. Chem.*, **235**, 1142 (1960)
26. Edmonds, M. and Abrams, R. J. *biol. Chem.*, **238**, PC 1186 (1963)
27. Engleman, D. M. and Moore, P. B. *Proc. natn Acad. Sci. U.S.A.*, **69**, 1997–1999 (1972)
28. Englander, S. W., Downer, N. W. and Teitelbaum, H. *A. Rev. Biochem.*, **41**, 903–924 (1972)
29. Epstein, W. and Beckwith, J. R. *A. Rev. Biochem.*, **37**, 411–436 (1968)
30. Freese, E. *Proc. natn Acad. Sci. U.S.A.*, **45**, 622–633 (1959)
31. Gallo, R. C., Yang, S. S. and Ting, R. C. *Nature, Lond.*, **228**, 927 (1970)
32. Gellert, M., Little, J. W., Oshinsky, C. K. and Zimmerman, S. B. *Cold Spring Harb. Symp. quant. Biol.*, **33**, 21–26 (1968)
33. Gilbert, W. and Dressler, D. *Cold Spring Harb. Symp. quant. Biol.*, **33**, 473–484 (1968)
34. Goulian, M., Kornberg, A. and Sinsheimer, R. L. *Proc. natn Acad. Sci. U.S.A.*, **58**, 2321–2328 (1967)
35. Greenleaf, A. L., Linn, T. G. and Losick, R. *Proc. natn Acad. Sci. U.S.A.*, **70**, 490–494 (1973)
36. Gros, F., Naono, S., Woese, C., Willson, C. and Attardi, G. In *Informational Macromolecules*, p. 387 (ed. by Vogel, H. J., Bryson, V. and Lampen, J. O.), Academic Press, New York (1963)
37. Grunberg–Manago, M. and Ochoa, S. *J. Am. chem. Soc.*, **77**, 3165–3166 (1955)
38. Hadjivassiliou, A. and Brawerman, G. *J. molec. Biol.*, **20**, 1–7 (1966)
39. Hall, B. D. and Spiegelman, S. *Proc. natn Acad. Sci. U.S.A.*, **47**, 137–146 (1961)
40. Hanson, J. and Huxley, H. E. *Nature, Lond.*, **173**, 973–976 (1954)
41. Harrington, W. F. *Proc. natn Acad. Sci. U.S.A.*, **68**, 685–689 (1971)
42. Hayton, G. J., Pearson, C. K., Scaife, J. R. and Keir, H. M. *Biochem. J.*, **131**, 499–508 (1973)
43. Helinski, D. R. and Clewell, D. B. *A. Rev. Biochem.*, **40**, 899–942 (1971)
44. Hershey, A. D., Dixon, J. and Chase, M. *J. gen. Physiol.*, **36**, 777–789 (1953)
45. Hirose, S., Okazaki, R. and Tamanoi, F. *J. molec. Biol.*, **77**, 501–517 (1973)
46. Holley, R. W., Apgar, J., Everett, G. A., Madison, J. T., Marquisee, M., Merrill, S. H., Penswick, J. R. and Zamir, A. *Science*, **147**, 1462–1465 (1965)
47. Huebner, J. R. and Todaro, G. J. *Proc. natn Acad. Sci. U.S.A.*, **64**, 1087–1094 (1969)
48. Huxley, H. E. *Science*, **164**, 1356–1366 (1969)
49. Ichimura, M. and Tsukada, K. *J. Biochem., Tokyo*, **69**, 823–828 (1971)
50. Ishitsura, H. and Kaji, A. *Proc. natn Acad. Sci. U.S.A.*, **66**, 168–173 (1970)
51. Jacob, F. and Monod, J. *J. molec Biol.*, **3**, 318–356 (1961)
52. Jacob, F., Ullman, A. and Monod, J. *C. R. acad. Sci.*, **258**, 3125–3128 (1964)
53. Jacob, F. and Wollman, E. L. *C. R. acad. Sci.*, **247**, 154–156 (1958)
54. Jacob, F. and Wollman, E. L. *Sexuality and the Genetics of Bacteria*, Academic Press, New York (1961)
55. Josse, J., Kaiser, A. D. and Kornberg, A. *J. biol. Chem.*, **236**, 864–875 (1961)
56. Kacian, D. L., Spiegelman, S., Bank, A., Terada, M., Metafora, S., Dow, L. and Marks, P. A. *Nature, Lond., New Biol.*, **235**, 167–169 (1972)

57. Kaempfer, R. *Nature, Lond.,* **228**, 534–537 (1970)
58. Kasamatsu, H., Robberson, D. L. and Vinograd, J. *Proc. natn Acad. Sci. U.S.A.,* **68**, 2252–2257 (1971)
59. Kim, S. H., Quigley, G. J., Suddath, F. L., McPherson, A., Sneden, D., Kim, J. J., Weinzierl, J. and Rich, A. *Science,* **179**, 285–288 (1973)
60. Kit, S. J. *J. molec. Biol.,* **3**, 711–716 (1961)
61. Kornberg, A. *Science,* **163**, 1410–1418 (1969)
62. Kunitz, M. *J. gen. Physiol.,* **24**, 15 (1940)
63. Kwan, S.-W. and Brawerman, G. *Proc. natn Acad. Sci. U.S.A.,* **69**, 3247–3250 (1972)
64. Langridge, R., Seeds, W. E., Wilson, H. R., Hooper, C. W., Wilkins, M. H. F. and Hamilton, L. D. *J. biophys. biochem. Cytol.,* **3**, 767–778 (1957)
65. Lederberg, J. *Physiol. Rev.,* **32**, 403–430 (1952)
66. Lengyel, P. and Söll, D. *Bacteriol. Rev.,* **33**, 264–301 (1969)
67. Levinthal, C. *Genetics,* **39**, 169 (1954)
68. Linderstrøm-Lang, K. *Proteins and Enzymes: Lane Lectures no. 6*, p. 93. University Press, Stanford, Calif. (1952)
69. Lucas-Lenard, J. and Lipmann, F. *A. Rev. Biochem.,* **40**, 409–448 (1971)
70. Lukanidin, E. M., Zalmanzon, E. S., Komaromi, L., Samarina, O. P. and Georgiev, G. P. *Nature, Lond., New Biol.,* **238**, 193–197 (1972)
71. McLaughlin, C. S., Warner, J. R., Edmonds, M., Nakazato, H. and Vaughan, M. H. *J. biol. Chem.,* **248**, 1466–1471 (1973)
72. Maas, W. K. and Clark, A. J. *J. molec. Biol.,* **8**, 365–370 (1964)
73. Mainwaring, W. I. P. and Peterken, B. M. *Biochem. J.,* **125**, 285–295 (1971)
74. Marcker, K. A. and Sanger, F. *J. molec. Biol.,* **8**, 835–840 (1964)
75. Marmur, J. and Doty, P. *J. molec. Biol.,* **3**, 585–594 (1961)
76. Marmur, J. and Lane, D. *Proc. natn Acad. Sci. U.S.A.,* **46**, 453–461 (1960)
77. Marvin, D. A., Spencer, M. and Wilkins, M. H. F. *Nature, Lond.,* **182**, 387–388 (1958)
78. Masters, M. and Broda, P. *Nature, Lond., New Biol.,* **232**, 137–140 (1971)
79. Meselson, M. and Stahl, F. W. *Proc. natn Acad. Sci. U.S.A.,* **44**, 671–682 (1958)
80. Min Jou, W., Haegeman, G., Usebaert, M. and Fiers, W. *Nature, Lond.,* **237**, 82–88 (1972)
81. Monod, J., Jacob, F. and Gros, F. *Biochem. Soc. Symp.,* **21**, 104–132 (1961)
82. Moore, P. B., Huxley, H. E. and Derosier, D. J. *J. molec. Biol.,* **50**, 279–295 (1970)
83. Nirenberg, M. W. and Mattaei, J. H. *Proc. natn Acad. Sci. U.S.A.,* **47**, 1580–1588 (1961)
84. Nirenberg, M. W. and Matthaei, J. H. *Proc. natn Acad. Sci. U.S.A.,* **47**, 1588–1602 (1961)
85. Nishizuka, Y. and Lipmann, F. *Proc. natn Acad. Sci. U.S.A.,* **55**, 212–219 (1966)
86. Okazaki, R., Okazaki, T., Sakabe, K., Sugimoto, K., Kainuma, R., Sugino, A. and Iwatsuki, N. *Cold Spring Harb. Symp. quant. Biol.,* **33**, 129–142 (1968)
87. Palade, G. E. and Siekevitz, P. *J. biophys. biochem. Cytol.,* **2**, 171–198, 671–690 (1956)
88. Pauling, L. and Corey, R. B. *Proc. natn Acad. Sci. U.S.A.,* **37**, 235–240, 261–271 (1951)

89. Pauling, L. and Corey, R. B. *Proc. natn Acad. Sci. U.S.A.*, **37**, 251–256 (1951)
90. Perlman, S., Abelson, H. T. and Penman, S. *Proc. natn Acad. Sci. U.S.A.*, **70**, 350–353 (1973)
91. Perutz, M. F. *Nature, Lond.*, **228**, 726–734, 734–739 (1970)
92. Perutz, M. F., Bolton, W., Diamond, R., Muirhead, H. and Watson, H. C. *Nature, Lond.*, **203**, 687–690 (1964)
93. Prichard, P. M., Gilbert, J. M., Shafritz, D. A. and Anderson, W. F. *Nature, Lond.*, **226**, 511–514 (1970)
94. Printz, M. P. and Hippel, P. H. von *Proc. natn Acad. Sci. U.S.A.*, **53**, 363–370 (1965)
95. Radloff, R., Bauer, W. and Vinograd, J. *Proc. natn Acad. Sci. U.S.A.*, **57**, 1514–1521 (1967)
96. Ramachandran, G. N. and Kartha, G. *Nature, Lond.*, **176**, 593–595 (1955)
97. Riddle, D. L. and Roth, J. R. *J. molec. Biol.*, **66**, 495–506 (1972)
98. Robberson, D. L., Kasamatsu, H. and Vinograd, J. *Proc. natn Acad. Sci. U.S.A.*, **69**, 737–741 (1972)
99. Roberts, J. W. *Cold Spring Harb. Symp. quant Biol.*, **35**, 121–126 (1970)
100. Robertson, H. D., Barrell, B. G., Weith, H. L. and Donelson, J. E. *Nature, Lond., New Biol.*, **241**, 38–40 (1973)
101. Roeder, R. G. and Rutter, W. J. *Proc. natn Acad. Sci. U.S.A.*, **65**, 675–682 (1970)
102. Roodyn, D. B. and Wilkie, D. *The Biogenesis of Mitochondria*, Methuen, London (1968)
103. Rosenberg, J. M., Seeman, N. C., Kim, J. J. P., Suddath, F. L., Nicholas, H. B. and Rich, A. *Nature, Lond.*, **243**, 150–154 (1973)
104. Ross, J., Aviv, H., Scolnick, E. and Leder, P. *Proc. natn Acad. Sci. U.S.A.*, **69**, 264–268 (1972)
105. Sanger, F. and Thompson, E. O. P. *Biochem. J.*, **53**, 353–374 (1953)
106. Schäfer, R. and Zillig, W. *Eur. J. Biochem.*, **33**, 201–206 (1973)
107. Setlow, P., Brutlag, D. and Kornberg, A. *J. biol. Chem.*, **247**, 224–231 (1972)
108. Setlow, P. and Kornberg, A. *J. biol. Chem.*, **247**, 231–240 (1972)
109. Sheiness, D. and Darnell, J. E. *Nature, Lond., New Biol.*, **241**, 265–268 (1973)
110. Sinsheimer, R. L. *J. molec. Biol.*, **1**, 43–53 (1959)
111. Sirlin, J. L. *Progr. Biophys. biophys. Chem.*, **12**, 25–66 (1962)
112. Smith, A. E. and Marcker, K. A. *Nature, Lond.*, **226**, 607–610 (1970)
113. Smithies, O., Connell, G. E. and Dixon, G. H. *Nature, Lond.*, **196**, 232–236 (1962)
114. Sodek, J., Hodges, R. S., Smillie, L. B. and Jurasek, L. *Proc. natn Acad. Sci. U.S.A.*, **69**, 3800–3804 (1972)
115. Spiegelman, W. G., Reichardt, L. F., Yaniv, M., Heinemann, S. F., Kaiser, A. D. and Eissen, H. *Proc. natn Acad. Sci. U.S.A.*, **69**, 3156–3160 (1972)
116. Spirin, A. S. *Eur. J. Biochem.*, **10**, 20–35 (1969)
117. Stanier, R. Y. and van Niel, C. B. *Arch. Mikrobiol.*, **42**, 17–35 (1962)
118. Sugino, A., Hirose, S. and Okazaki, R. *Proc. natn Acad. Sci. U.S.A.*, **69**, 1863–1867 (1972)
119. Suzuki, E., Crewther, W. G., Fraser, R. D. B., Macrae, T. P. and McKean, N. M. *J. molec. Biol.*, **73**, 275–278 (1973)
120. Szent-Gyorgyi, A. G., Cohen, C. and Kendrick-Jones, J. *J. molec. Biol.*, **56**, 239–258 (1971)

121. Szybalski, W., Bøure, K., Fiandt, M., Hayes, S., Hradecna, Z., Kumar, S., Lozeron, H. A., Nijkamp, H. J. J. and Stevens, W. F. *Cold Spring Harb. Symp. quant. Biol.* **35**, 341–353 (1970)
122. Takanami, M., Okamoto, T. and Sugiura, M. *J. molec. Biol.*, **62**, 81–88 (1972)
123. Temin, H. M. *Virology*, **23**, 486 (1964)
124. Temin, H. M. *Cancer Res.*, **28**, 1835 (1968)
125. Temin, H. M. *Perspect. Biol. Med.*, **14**, 11–26 (1970)
126. Thompson, E. B., Granner, D. K. and Tomkins, G. M. *J. molec. Biol.*, **54**, 159–175 (1970)
127. Tomkins, G. M. and Martin, D. W. Jr *A. Rev. Genetics*, **4**, 91–106 (1970)
128. Traub, W., Yonath, A. and Segal, D. M. *Nature, Lond.*, **221**, 914–917 (1969)
129. Vallee, B. L. and Williams, R. J. P. *Proc. natn Acad. Sci. U.S.A.*, **59**, 498–505 (1968)
130. Verma, I. M., Temple, G. F., Fan, H. and Baltimore, D. *Nature, Lond., New Biol.*, **235**, 163–167 (1972)
131. Vinograd, J., Lebowitz, J. and Watson, R. *J. molec Biol.*, **33**, 173–197 (1968)
132. Volkin, E. In *Molecular Genetics*, Part 1, p. 271. (ed. by Taylor, H. J.), Adademic Press, New York (1963)
133. Warner, J. R., Knopf, P. M. and Rich, A. *Proc. natn Acad. Sci. U.S.A.*, **49**, 122–129 (1963)
134. Watson, J. D. and Crick, F. H. C. *Nature, Lond.*, **171**, 737 (1953)
135. Weintraub, H. *Nature, Lond.*, **240**, 449–453 (1972)
136. Wetmur, J. G. and Davidson, N. *J. molec. Biol.*, **31**, 349–370 (1968)
137. Wolstenholme, D. *J. cell Biol.*, **56**, 247–255 (1973)
138. Ziff, E. B., Sedat, J. W. and Galibert, F. *Nature, Lond., New Biol.*, **241**, 34–38 (1973)
139. Zimm, B. H. *J. chem. Phys.*, **16**, 1093–1116 (1948)